Topics in Elementary Geometry

Second Edition

O. Bottema
(deceased)

Topics in Elementary Geometry

Second Edition

With a Foreword by Robin Hartshorne

Translated from the Dutch by Reinie Erné

 Springer

O. Bottema
(deceased)

Translator:
Reinie Erné
Leiden, The Netherlands
erne@math.leidenuniv.nl

ISBN: 978-0-387-78130-3   e-ISBN: 978-0-387-78131-0
DOI: 10.1007/978-0-387-78131-0

Library of Congress Control Number: 2008931335

Mathematics Subject Classification (2000): 51-xx

This current edition is a translation of the Second Dutch Edition of, *Hoofdstukken uit de Elementaire Meetkunde*, published by Epsilon-Uitgaven, 1987.

Printed on acid-free paper

9 8 7 6 5 4 3 2 1

springer.com

At school I was good in mathematics; now I discovered that
I found the so-called higher mathematics – differential
and integral calculus – easier than the complicated (but elementary)
plane geometry.

Hendrik B.G. Casimir,
Het toeval van de werkelijkheid,
Een halve eeuw natuurkunde,
Meulenhoff Informatief bv,
Amsterdam, 1983, p. 76.

Foreword

Oene Bottema (1901–1992) may not be so well known abroad, but in his own country he is "the great geometer". He graduated from the University of Groningen in 1924 and obtained his doctor's degree from Leiden University in 1927. He spent his early years as a high school teacher and administrator. He published extensively, and as his ability became known, he was made professor at the Technical University of Delft in 1941, and later rector of that university (1951–1959). With his encyclopedic knowledge of 19th-century geometry and his training in 20th-century rigor, he was able to make many contributions to elementary geometry, even as that subject was eclipsed by the modern emphasis on abstract mathematical structures. He also had a fruitful collaboration with engineers and made substantial contributions to kinematics, culminating in the book *Theoretical Kinematics*, with Bernard Roth, in 1979. Throughout his life he was inspired by geometry and poetry, and favored elegant succinct proofs.

This little book, first published in 1944, then in a second expanded edition in 1987, gives us a glimpse into his way of thinking. It is a series of vignettes, each crafted with elegance and economy. See, for example, his proof of the Pythagorean theorem (1.2), which requires only one additional line to be drawn. And who can imagine a simpler proof of the nine-point circle (4.1)? There is ample coverage of the modern geometry of the triangle: the Simson line, Morley's theorem, isogonal conjugates, the symmedian point, and so forth. I was particularly struck by the proof of the concurrence of the altitudes of a triangle, independent of the parallel postulate (3.1). This book has many gems to delight both the novice and the more experienced lover of geometry.

November 2007

Robin Hartshorne
Berkeley, CA

Preface to the Second Edition

This is the second, revised and supplemented, edition of a book with the same title that appeared in 1944. It was part of the Mathematics Section of the series known as *Servire's Encyclopaedie*. It was written during the oppressive reality of occupation, darkness, and sadness. For the author, it meant that during a few open hours he could allow himself to exchange the worries of daily life for the pleasure that comes from fine mathematical figures and the succession of syllogisms ("it follows that", "therefore", "hence"') crowned with a well-earned *quod erat demonstrandum*. In those hours he encountered numerous historical figures, starting with the legendary PYTHAGORAS and meeting, among many others, PTOLEMY, TORRICELLI (of the barometer), and the hero EULER (not a true geometer: for him, a geometric figure was a call to exercise his desire to compute); continuing through the heavily populated nineteenth century, appropriately named the golden age of geometry; along the recent past of MORLEY's triangle to some recent theorems.

What lies before you can best be called an anthology of geometric truths, a subjective choice determined by the personal preferences of the author. The few principles that led the anthologist had a negative character: restriction to the plane, refraining from axiomatic buildup, avoidance of problems of constructive nature.

This document could not have been made without the sympathy of a number of good friends and the gratefully accepted help with the readying of the final text for publication.

In the first place, however, the gratitude of the author goes out to the editor of *Epsilon Uitgaven* whose suggestion of reediting a text that was more than forty years old was accepted with understandable satisfaction.

O.B.

Contents

1

The Pythagorean Theorem

1.1

The Pythagorean theorem is not only one of the most important and oldest theorems in our geometry, it is also very well known and you might even say popular. This is due to its simplicity, which nonetheless does not imply that its proof is obvious. The theorem states that, *for a right triangle with legs of length a and b and hypotenuse of length c, we have the relation $a^2 + b^2 = c^2$.* In more geometric terms, this statement becomes: the area of a square having the hypotenuse as a side is equal to the sum of the areas of the two squares, each of which has one of the other legs of the triangle as a side. The theorem takes its name from the semi-legendary philosopher PYTHAGORAS of SAMOS who lived around 500 BC [Hea2]. Whether he was aware of the theorem, and if so, whether he and his followers had a proof of it, cannot be said with certainty. However, it appears that already many centuries prior to this, the Babylonians were familiar with the theorem. Throughout the ages, many different proofs and many different types of proofs have been given for the theorem; we present four of them here.

1.2

Consider the following three properties: (1) corresponding distances in two similar figures are proportional, (2) the area of a figure can be viewed as the sum of areas of triangles; for a curved figure this should be replaced by the limit of a sum, and (3) said succinctly, the area of a triangle is equal to the product of two lengths. It follows from these that the areas of two similar figures are proportional to the *squares* of the corresponding lengths. This implies that the Pythagorean theorem can be stated in a broader sense: *if on the sides of a right triangle, we set three mutually similar figures with areas O_a, O_b, and O_c, then $O_a + O_b = O_c$.* Conversely, a proof of the theorem can be given by showing that we can find a set of mutually similar figures set on

O. Bottema, *Topics in Elementary Geometry*,
DOI: 10.1007/978-0-387-78131-0_1, © Springer Science+Business Media, LLC 2008

the legs and the hypotenuse of a right triangle, such that the sum of the areas
of the first two is equal to the area of the third. In fact, such a set immediately
appears by drawing the altitude CD; the triangles CAB, DAC, and DCB
are similar, because they have congruent angles (Fig. 1.1).

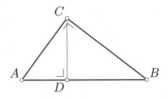

Fig. 1.1.

1.3

EUCLID's *Elements* (ca. 300 BC) [Hea1], that great monument of Greek culture, contains a systematic exposition and rigorous logical construction of the
mathematical properties of the plane and of 3-space. Until the appearance of
the modern criticism of the second half of the nineteenth century, it remained
a faultless classical example of a mathematical argumentation. Supposed im-
perfections found by earlier generations often turned out to be misunderstood
qualities. Our theorem can be found in the 1st Book as Proposition 47. It is
proved as follows (Fig. 1.2):

   We draw the extended altitude CDK, followed by AE, BF, CG, and CH.
The triangles FAB and CAG are congruent because they have two equal
sides with congruent included angle. The first one has the same area as triangle FAC, which has the same base FA and the same height, and the second
one has the same area as triangle KAG, for a similar reason. It follows that
square $ACA'F$ has the same area as rectangle $ADKG$. Likewise, the area of
square $BCB'E$ is equal to the area of rectangle $BDKH$.

1.4

If $a \geq b$ and we set the triangle in a square with side c in four positions
obtained from one another by rotation over a right angle about the center of
the square (Fig. 1.3), the four figures do not overlap and fill the square up to
a square with sides of length the difference $a - b$. We therefore have

$$c^2 = 4 \times \tfrac{1}{2} \times ab + (a - b)^2 ,$$

or, after simplifying,

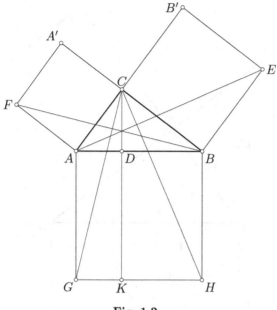

Fig. 1.2.

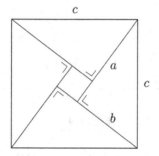

Fig. 1.3.

$$c^2 = a^2 + b^2 .$$

This proof can be found in the work of the ancient Indian mathematician BHASKARA (ca. 1100) [Bha], [Col], but according to the historian CANTOR, it was already known to the Indians some centuries before Christ [Can], see also [Gard]

1.5

The points E, C, and F (Fig. 1.4) seem to lie on one line, the external bisector

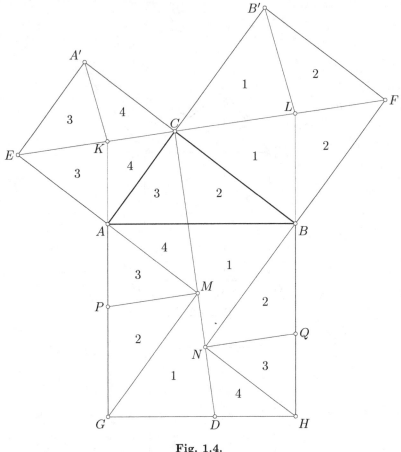

Fig. 1.4.

of angle C. We also draw the internal bisector CD, which lies perpendicular to it, extend GA to K, HB to L, EA to M, and FB to N, and draw MP and NQ perpendicular to CD.

Simple arguments with congruent triangles, which nonetheless need to be worked out with the necessary care, show that triangles indicated with the same number in the figure have equal areas. The correctness of the theorem immediately follows. The figure is all the more remarkable because triangle ABC itself is divided into pieces that are congruent to parts 2 and 3, respectively. The areas of triangles 1, 2, 3, and 4 are, respectively,

$$\frac{a^3}{2(a+b)}, \quad \frac{a^2 b}{2(a+b)}, \quad \frac{ab^2}{2(a+b)}, \quad \text{and} \quad \frac{b^3}{2(a+b)}.$$

This proof was given by EPSTEIN in 1906. Such dissection proofs are not new, another well-known example is the proof of THÁBIT IBN QURRA, which was written up by AN-NAIRIZI (ca. 900), an Arabic commentator on the Elements [Bes]. When $a < b$, the square a^2 is divided into two triangles and a quadrilateral, and the square b^2 into a triangle and a quadrilateral, after which we only need to move the five pieces around to obtain a cover of the square c^2.

1.6

The number of proofs of the Pythagorean theorem, including some by men who are known outside of mathematics, like the writer MULTATULI [Mul] (Idea 529), and President GARFIELD of the U.S.A. [Garf], is so great that collections of such proofs have been published. It is therefore no coincidence that this theorem was taken as the starting point for attempts to compare proofs of a given geometric theorem with each other in order to establish objective criteria to measure the *simplicity* of a proof. For example, the number of times that existing theorems are used could be taken as a measure. However, such observations were found to be of little interest, probably due to the arbitrariness of the chosen criteria. Moreover, the question of which proof is the simplest seems to be more one of psychological than one of mathematical nature. Similar observations concerning the simplicity of geometric *constructions* for a given figure have found more support. Although the same objections still hold, there is less arbitrariness because of the limitations imposed by the instruments that are allowed, the ruler and compass. Moreover, in the case of technical drawings, economic considerations may also play a role. This may be an explanation for the attention given to the *geometrography* of LEMOINE (1888) [Lem].

1.7

The figure of a right triangle with three squares set on its sides is so well known that at one point people proposed to use it to begin an interplanetary dialog. Surely a culture that is somewhat similar to ours, even on another planet, would know the Pythagorean theorem. The suggestion was to draw the figure on a huge scale somewhere on earth, either through illuminated signs, or as a crop formation. The nature of the first answer from outer space would then be awaited in suspense, and with the necessary patience.

1.8

Without being detrimental to the extraordinary importance of our theorem, we must note that its importance for elementary geometry is not due to its having a central place in the geometric system, but to its usefulness in computing lengths and therefore angles, areas, and volumes. Moreover, "theorems" that find their origin in the Pythagorean theorem are often no more than results that give these computations in a general form. In trigonometry, and in elementary analytic geometry, it occupies a fundamental position. However, its significance is limited to the normal, *Euclidean* geometry; in geometric systems constructed as an extension of or as an alternative to this one, the theorem does not hold. In projective, affine, and conformal geometry, it does not even make sense to question its validity; in non-Euclidean geometry such as the geometry of the sphere, the theorem does not hold since it relies on the existence of similar figures. We will not go into such questions here, just as we will not go into those posed by differential geometry concerning the validity of the theorem, or of a generalization thereof, for *small* triangles.

2

Ceva's Theorem

2.1

A triangle has *four* sets of classical special lines: the *angle bisectors d*, the *medians z*, the *altitudes h*, and the *perpendicular bisectors m*. They are special because in any triangle, the three lines of one type concur at a point, which is then called a *special point* of the triangle. The proofs of the four concurrences can be given in different ways. For the moment, let us not consider the perpendicular bisectors m because unlike d, z, and h, they are not *cevians* of the triangle as they do not each pass through a vertex.

2.2

Of the other three, the proof for the angle bisectors d is the simplest, by using the notion of *locus*. In triangle ABC, the bisector d_a of angle A contains all points that are equidistant from AB and AC, and no other points. As d_b has the analogous property, their intersection is equidistant from AB, BC, and CA, and therefore is a point of d_c. It is clear that this argument can immediately be extended to three *arbitrary* concurrent cevians. After all (Fig. 2.1), if t is an arbitrary line through the intersection point S of two lines l and m, P_1 and P_2 are two points on t, Q_1 and Q_2 are their projections

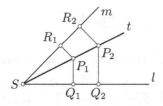

Fig. 2.1.

O. Bottema, *Topics in Elementary Geometry*,
DOI: 10.1007/978-0-387-78131-0_2, © Springer Science+Business Media, LLC 2008

on l, and R_1 and R_2 are those on m, then it follows from simple considerations of proportionality that $P_1 Q_1 : P_1 R_1 = P_2 Q_2 : P_2 R_2$. Consequently, for the points on a cevian t_1 going through the vertex A of triangle ABC, the distances from AB and AC are in a fixed ratio v_1. Moreover, if we restrict ourselves to points inside the triangle, or at least in the interior of angle BAC, then such a cevian is also the locus of the points with this property.

For t_2 going through B, let v_2 be the ratio of the distances from a point on t_2 to BC and BA. For t_3 going through C, let v_3 be the ratio of the distances from a point on t_3 to CA and CB. Then the intersection point of t_1 and t_2 will have distances from AC, BA, and CB that are in the ratio $1 : v_1 : v_1 v_2$, so that this intersection point lies on t_3 if $1 : v_1 v_2 = v_3$, that is, $v_1 v_2 v_3 = 1$. *The necessary and sufficient condition for the concurrence of the cevians t_1, t_2, and t_3 is therefore that $v_1 v_2 v_3 = 1$.*

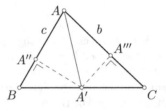

Fig. 2.2.

If t_1 is the median AA' from A (Fig. 2.2), and A'' and A''' are the projections of A' on AB and AC, then the altitudes $A'A''$ and $A'A'''$ of triangles $AA'B$ and $AA'C$ are inversely proportional to the bases AB and AC, so that $v_1 = b/c$. Likewise, $v_2 = c/a$ and $v_3 = a/b$, so that the desired relation holds. The medians z are therefore concurrent.

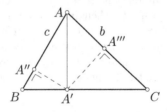

Fig. 2.3.

If t_1 is the altitude AA' (Fig. 2.3) of the acute triangle ABC, then

$$v_1 = \frac{A'A''}{A'A'''} = \frac{AA' \cos B}{AA' \cos C} = \frac{\cos B}{\cos C}.$$

It now follows from

$$v_2 = \frac{\cos C}{\cos A} \quad \text{and} \quad v_3 = \frac{\cos A}{\cos B}$$

that the altitudes h are concurrent.

2.3

We have characterized each cevian using the ratio v of the distances from one of its points to two sides. We can also characterize t_1 using the ratio u_1 of the areas of the triangles ABA' and CAA', or, which is the same since the triangles have equal altitudes, using the ratio of the segments $A'B$ and $A'C$ into which t_1 divides the opposite side. We then have $u_1 = (c/b)v_1$ and so on, so that *the necessary and sufficient condition that these ratios must fulfill for the cevians to be concurrent is that $u_1u_2u_3 = 1$*. In this form, the theorem is known as CEVA's theorem (1678) [Cev].

2.4

The theory presented above holds only if we restrict ourselves to cevians that intersect the triangle. If we also allow cevians that meet an *extension* of the opposite side, the proof no longer holds, because in that case, the locus of the points with given v_1 consists of *two* lines through A. Such problems can be solved by agreeing, as was first done systematically by MÖBIUS (1827) [Möb], the distance from a point to a line, and in particular opposite signs to points lying on opposite sides of a line. We will call the distance from a point to a side of the triangle positive if the point lies on the same side as the third vertex. We continue to consider the lengths of the sides a, b, and c to be positive. Consequently, keeping the relations of type $u_1 = (c/b)v_1$ leads to the attribution of a sign to each of the u_i. For example, u_1, the ratio of the distances BA' and $A'C$, will be positive or negative according to whether A' lies on side BC or on its extension.

Our notation reflects this agreement, as we will consider the distance PQ to be the opposite of the distance QP. Let us remark that it only makes sense to attribute a sign to the distance between two points if we compare distances on a given line (or possibly on two parallel lines). If, as is the case for CEVA's theorem, we are interested in the ratio of two distances BA' and $A'C$ on one side of a triangle, then, strictly speaking, it is not necessary to give each individual distance a sign. We can, however, agree to call BA' positive if A' lies on the same side of B as C; $A'C$ must then be called positive if A' lies on the same side of C as B.

With these conventions, where v can also be negative, the ratio v always corresponds to a unique cevian. The same holds for u, on condition that we handle the case $u = -1$ with the necessary care. CEVA's theorem holds without modification, by which we mean that if the relation holds, and two of the cevians intersect, then the third also passes through their intersection point. However, it can now also occur that two cevians t_1 and t_2 are parallel. In that case, if $u_1 u_2 u_3 = 1$ then t_3 must also be parallel to t_1 and t_2, as can be shown by contradiction. If we moreover extend the expression "to be concurrent" to include "to be parallel", a decision which, if not justified on other grounds, is justified on terminological grounds, then we can leave the wording of CEVA's theorem unchanged. In the case of parallel cevians, the following (non-independent) relations hold:

$$u_2 + 1 = -\frac{1}{u_1}, \qquad u_3 + 1 = -\frac{1}{u_2}, \qquad \text{and} \qquad u_1 + 1 = -\frac{1}{u_3}.$$

These, of course, imply that $u_1 u_2 u_3 = 1$.

2.5

We can now complete the proof of the concurrence of the *altitudes*, because even in obtuse triangles the equalities

$$v_1 = \frac{\cos B}{\cos C}$$

and so on hold because of the sign conventions in trigonometry. Of course, altitudes cannot be parallel.

The concurrence of a set of cevians composed of *one interior and two exterior angle bisectors* can be shown in a similar way. For example, we have $u_1 = c/b$, $u_2 = -a/c$, and $u_3 = -b/a$. As $u_3 + 1 + 1/u_2 = (a - b - c)/a \neq 0$, the bisectors cannot be parallel.

2.6

CEVA's theorem can be very useful in showing the concurrence of cevians, even if they are of another type than those mentioned above. For example, for the cevians AE, BF, and CD, it follows from Fig. 1.2 (page 3) that $u_1 = a/b$, $u_2 = a/b$, and $u_3 = b^2/a^2$. Consequently, these three auxiliary lines are concurrent.

If the incircle of a triangle *touches its sides at A', B', and C'* (Fig. 2.4), then $AB' = AC'$ ($= p_1$), whence $u_1 = p_2/p_3$. The analogous equalities hold for u_2 and u_3. The lines AA', BB', and CC' therefore concur at a special point which, although CEVA already knew of its existence, is usually called the GERGONNE *point* after a later discoverer.

We obtain a completely analogous theorem if we replace the incircle by *one of the three excircles* (Fig. 2.5, where we have chosen the excircle tangent to BC).

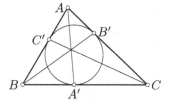

Fig. 2.4.

We have $AB' = AC' = q_1$, $BC' = BA' = q_2$, and $CA' = CB' = q_3$, whence $u_1 = q_2/q_3$, $u_2 = -q_3/q_1$, and $u_3 = -q_1/q_2$, which again implies that AA', BB', and CC' concur. The points where the four tritangent circles touch a triangle also give rise to concurrent cevians when grouped in another way. If we take *the three excircles* k_a (tangent to BC), k_b, and k_c, and let A', B', and C' be the points where they respectively touch BC, CA, and AB, then AA', BB', and CC' concur at a point named after NAGEL, often denoted by Na [Nag]. To prove this, we need to use generally known relations such as $BA' = s - c$. We can also take a combination of two excircles and the incircle, choosing the three tangent points carefully. In Chapter 25 we will come back to the relations between a number of things mentioned in this chapter. We will also give additional applications of CEVA's theorem.

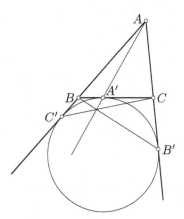

Fig. 2.5.

3

Perpendicular Bisectors; Concurrence

3.1

Let us consider the perpendicular bisectors m of a triangle. If we view m_a as being the locus of the points that are equidistant from B and C, do the same for m_b and m_c, and consider the intersection of two of these, the concurrence of these lines immediately follows.

Using this result, GAUSS gave a simple proof of the concurrence of the altitudes (Fig. 3.1). He constructed a triangle $A''B''C''$ circumscribing trian-

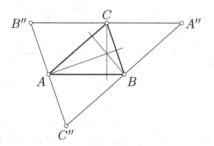

Fig. 3.1.

gle ABC, with sides parallel to those of the latter. It follows from the properties of a parallelogram that AB'' and AC'' are both equal to BC, so that the altitudes of ABC turn out to be the perpendicular bisectors of $A''B''C''$.

There exists another proof of this theorem. If triangle ABC is acute (Fig. 3.2) and AA', BB', and CC' are the altitudes, then the similarity of triangles $AB'C'$, $A'BC'$, and $A'B'C$ to ABC implies that these altitudes are the angle bisectors of the *pedal triangle* $A'B'C'$. For an obtuse triangle, we need to change the wording, because two of the altitudes are exterior angle bisectors and the third (from the vertex of the obtuse angle) is an interior angle

O. Bottema, *Topics in Elementary Geometry*,
DOI: 10.1007/978-0-387-78131-0_3, © Springer Science+Business Media, LLC 2008

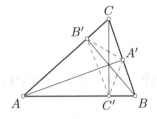

Fig. 3.2.

bisector of $A'B'C'$. It is remarkable that all three proofs of the concurrence of the altitudes outlined here use the *parallel postulate*, either directly or through properties of proportionality and similarity, while in non-Euclidean geometry, which differs from ours by not admitting this postulate, this concurrence theorem essentially still holds. The following proof, by GUDERMANN (1835), does not depend on the parallel postulate. Let AA' and BB' (Fig. 3.3) be two

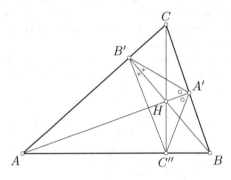

Fig. 3.3.

altitudes that meet at H. If $A'B'$ is reflected in AA' and in BB', giving rise to triangle $A'B'C''$, then H is the center of the incircle of that triangle. The points C (as the intersection point of two exterior angle bisectors) and A and B (as the intersection points of an interior and an exterior angle bisector) are the centers of the excircles. It follows from this that AB and CH both pass through C'' and that they are perpendicular to each other, as they are respectively the exterior and interior angle bisector of C''. This completes the proof.

This proof is incomplete in the sense that the existence of H and C'' is tacitly assumed. In a geometry that does not allow non-intersecting lines (the *elliptic geometry*, which is closely related to the geometry of the sphere), the proof is therefore correct; in the *hyperbolic geometry*, additional thought is required. In fact, the theorem does not hold in the usual form: the three altitudes do not necessarily have a "normal" intersection point (lying in the Euclidean plane).

3.2

The argumentation given above for the lines m can be transferred to the figure of three arbitrary concurrent lines perpendicular to the three sides of the triangle. A line from a point A', perpendicular to BC, for example, is the locus of the points P for which $PB^2 - PC^2$ is fixed. The proof follows from the Pythagorean theorem and from the fact that the point A' on BC is completely determined by the expression $A'B^2 - A'C^2$. By an argument analogous to that of Section 2.2, it now follows that: *a necessary and sufficient condition for the concurrence of the perpendiculars from P, Q, and R on respectively BC, CA, and AB is that $BP^2 - PC^2 + CQ^2 - QA^2 + AR^2 - RB^2 = 0$.*

From this theorem also follows a simple proof of the concurrence of the altitudes. Let us give another application (Fig. 3.4). If we draw triangles $A'BC$,

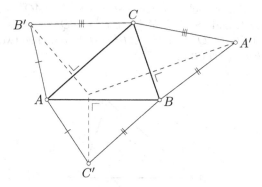

Fig. 3.4.

$B'AC$, and $C'AB$ outward on the sides of triangle ABC, in such a way that $AB' = AC'$, $BC' = BA'$, and $CA' = CB'$, then the altitudes from A', B', and C' are concurrent. In this figure, which is well known in solid geometry, we can recognize the net of a (not necessarily regular) tetrahedron.

3.3

Let us denote the circle with center M and radius R by $M(R)$. Given a point P, for an arbitrary line l through P, which intersects the circle $M(R)$ at A and B, we consider the expression $PA \times PB$. This turns out to be independent of l. It is called the *power* of P with respect to the circle. It is attributed a sign according to our earlier conventions, so that the power of a point inside the circle is negative, and is apparently equal to $PM^2 - R^2$. If we have two non-concentric circles $M_1(R_1)$ and $M_2(R_2)$, then according to 3.2, the locus of the points of equal power with respect to these circles is *a line perpendicular to* M_1M_2, passing through the point M_{12} for which

$$M_1M_{12}^2 - M_{12}M_2^2 = R_1^2 - R_2^2 \,.$$

For two circles that intersect, this *radical axis*, or *radical line*, is equal to the line joining the intersection points. Our theorem now immediately leads to the conclusion that for three circles whose centers are the vertices of a triangle, *the three radical axes concur*, at a point called the *radical center* (or *power center*) of the three circles.

3.4

If ABC and PQR are triangles such that the perpendiculars from A on QR, from B on RP, and from C on PQ are concurrent at a point S, then the perpendiculars from P on BC, from Q on CA, and from R on AB are also concurrent (Fig. 3.5). Namely, we have $SQ^2 - SR^2 = AQ^2 - AR^2$, $SR^2 - SP^2 = BR^2 - BP^2$, and $SP^2 - SQ^2 = CP^2 - CQ^2$, whence we obtain, by addition: $BP^2 - CP^2 + CQ^2 - AQ^2 + AR^2 - BR^2 = 0$. The two triangles are said to be *orthologic* (STEINER, 1827).

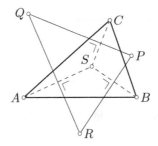

Fig. 3.5.

3.5

If P, Q, and R are the projections of A, B, and C on the line l (Fig. 3.6), then $PB^2 = PQ^2 + BQ^2$, $PC^2 = PR^2 + CR^2$, $QC^2 = QR^2 + CR^2$, $QA^2 = QP^2 + AP^2$, $RA^2 = RP^2 + AP^2$, $RB^2 = RQ^2 + BQ^2$, and therefore $PB^2 - PC^2 + QC^2 - QA^2 + RA^2 - RB^2 = 0$. The perpendiculars from P, Q, and R on BC, CA, and AB therefore concur at one point, which we call the *orthopole* of l with respect to the triangle ABC (NEUBERG, 1878 [Neu]).

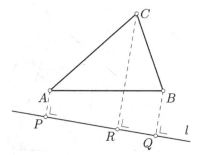

Fig. 3.6.

3.6

If we choose the points A', B', and C' on BC, CA, and AB, respectively, in such a way that the perpendiculars from these points concur at a point T, then $A'B'C'$ is called the *pedal triangle* of ABC for the point T. An interesting question is what restrictions exist on the position of T when we impose certain conditions on its pedal triangle. Even in simple cases, the answer often cannot be given within elementary geometry. If, for example, we impose that AA', BB', and CC' concur at a point S, then the locus of T, as well as that of S, turns out to be a *curve* meeting a line at three points (DARBOUX cubics [Dar] and LUCAS cubics [Luc1], [Luc2]). The situation is simpler if we add the condition that A', B', and C' are collinear; we will consider that case in Chapter 9.

3.7

Let us denote the area of any figure by the name of the figure enclosed in square brackets. An unusual result holds for the *area of an arbitrary pedal triangle* (Fig. 3.7). Let $A'B'C'$ be the pedal triangle for T and let S be the second intersection point of AT and the circumcircle $M(R)$. As $AB'TC'$ and $BA'TC'$ are cyclic quadrilaterals, we have $\angle B'C'T = \angle B'AT = \angle CBS$ and $\angle TC'A' = \angle TBA'$, so that $\angle C' = \angle B'AT + \angle TBA' = \angle TBS$. For the area $[A'B'C']$ of triangle $A'B'C'$, we therefore have

$$[A'B'C'] = \tfrac{1}{2}B'C' \times A'C' \times \sin C'$$
$$= \tfrac{1}{2}TA \sin A \times TB \sin B \times \sin \angle TBS$$
$$= \tfrac{1}{2}TA \times TS \times \sin A \times \sin B \times \sin C.$$

The product $TA \times TS$ is the power of T with respect to the circle, so that we find (GERGONNE, 1823)

$$[A'B'C'] = \tfrac{1}{2}(R^2 - MT^2)\sin A \sin B \sin C.$$

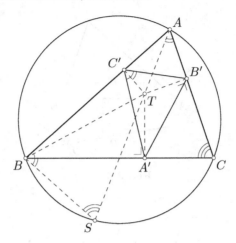

Fig. 3.7.

By convention, $[A'B'C']$ is viewed as being positive if $A'B'C'$ describes the same sense of rotation as ABC. Points T that are equidistant from M therefore have pedal triangles of equal area.

4

The Nine-Point Circle and Euler Line

4.1

Let $A'B'C'$ be the pedal triangle of triangle ABC for a point H (Fig. 4.1). Let M_a, M_b, and M_b be the midpoints of its sides. Then $A'M_aM_bM_c$ is a trapezoid that is isosceles, because $A'M_c$ and M_aM_b are both equal to half of AB.

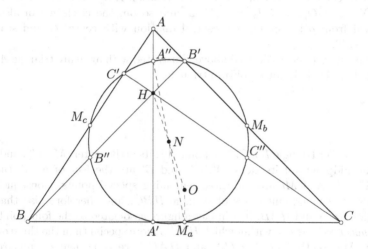

Fig. 4.1.

It therefore has a circumcircle n. In other words, the circle n through $M_aM_bM_c$ passes through A' and therefore also through B' and C'. Let A'' be the midpoint of AH, then M_cA'' is parallel to BB'. As M_cM_a is parallel to AC, M_aM_cA'' is a right angle. Consequently the circle with diameter M_aA'' passes through both A' and M_c, and therefore coincides with n. We thus obtain the

O. Bottema, *Topics in Elementary Geometry*,
DOI: 10.1007/978-0-387-78131-0_4, © Springer Science+Business Media, LLC 2008

following theorem: *the nine points M_a, M_b, M_c, A', B', C', A'', B'', and C'' lie on the same circle n, called the nine-point circle*. In 1765, EULER (1707–1783) showed that the first six points lie on one circle [Eul]). In 1820, BRIANCHON (1783–1864) and PONCELET (1778–1867) showed that this circle also contains the points A'', B'', and C'' [BP]. The name *nine-point circle* was introduced by TERQUEM in 1842 [Ter].

4.2

The points A'', B'', and C'' are obtained from A, B, and C by applying a central dilation with center H and scaling factor $1/2$. The circle n can therefore be constructed by applying this transformation to the circumcircle o of ABC. Let O be the center and R the radius of o, then the radius of n is equal to $R/2$, and the center N is the midpoint of HO. The latter also follows from the fact that M_a and A'' are diametrically opposite. But this in turn implies that $A''H = OM_a$, and therefore $AH = 2OM_a$. If the median AM_a meets the line HO at G, we have $AG = 2GO_a$, whence it follows that G is the centroid. We thus find that the classical special points H, G, and O lie on the same line (named after EULER) on which N also lies [Eul],[LH]. Their position with respect to each other is controlled by the relation $HN : NG : GO = 3 : 1 : 2$. As $GN = -GO/2$, $GM_a = -GA/2$, and so on, the circle n can also be obtained from o by applying a central dilation with center G and scaling factor $-1/2$.

All of the properties derived above, illustrated with an acute triangle ABC in Fig. 4.1, also hold for an obtuse triangle.

4.3

Let us consider triangle HBC. The point A is its orthocenter, M_a, C'', and B'' are the midpoints of its sides, A', C', and B' are the feet of its altitudes, and A'', M_c, and M_b are the three remaining special points. Consequently n is also the nine-point circle of triangle HBC, and therefore also that of triangles HCA and HAB. It is in fact a figure associated to the four points of an *orthocentric system*, within which H is no more special than the three other points. Apparently, HBC, HCA, and HAB have congruent circumcircles. Their EULER lines are AN, BN, and CN.

4.4

The theorem that states that n *touches the incircle internally and the excircles externally* is due to FEUERBACH (1822) [Feu]. We can easily prove this property through computations: we must in fact express the distance NI, respectively NI_a, in the given elements.

There also exists an interesting proof using *inversion*, which is discussed in Chapter 21.

5

The Taylor Circle

In Fig. 5.1, D, E, and F are the feet of the altitudes of triangle ABC. We consider the projections of these points on the remaining sides. Let D_2 and D_3 be the projections of D on, respectively, AC and AB, E_3 and E_1 those of E on BA and BC, and, finally, F_1 and F_2 those of F on CB and CA. This gives two new points on each side of the triangle. TAYLOR (1882) proved that *these six points lie on the same circle*, which was subsequently named after him. Let us give a proof using trigonometry. We have $AE = c \cos \alpha$

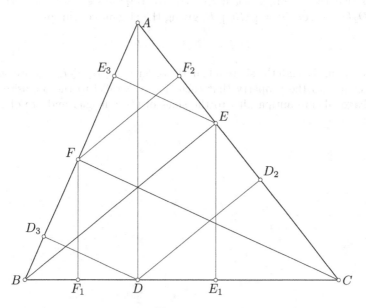

Fig. 5.1.

and $AE_3 = c \cos^2 \alpha$. Moreover, $BD_3 = BD \cos \beta = c \cos^2 \beta$. It follows that

O. Bottema, *Topics in Elementary Geometry*,
DOI: 10.1007/978-0-387-78131-0_5, © Springer Science+Business Media, LLC 2008

$AD_3 = c - BD_3 = c\sin^2\beta$, and therefore

$$AD_3 \times AE_3 = c^2\cos^2\alpha \times \sin^2\beta = 4R^2\cos^2\alpha\sin^2\beta\sin^2\gamma\,,$$

where R is the radius of the circumcircle of triangle ABC. To obtain $AD_2 \times AF_2$, we must interchange β and γ, which gives $AD_2 \times AF_2 = AD_3 \times AE_3$, with the well-known conclusion: D_2, D_3, E_3, and F_2 lie on a circle K_1. Likewise, E_3, E_1, F_1, and D_3 lie on a circle K_2. We can verify that the three vertices A, B, and C have equal powers with respect to the two circles. If K_1 and K_2 were distinct, the set of points of equal power with respect to the two circles would be a line, their radical axis. Here we have three non-collinear points with equal powers. The circles K_1 and K_2 therefore coincide, and consequently also coincide with K_3. The figure has more interesting properties. From $AF = b\cos\alpha$, it follows that $AF_2 = b\cos^2\alpha$. As $AE_3 = c\cos^2\alpha$, we have $AE_3 : AF_2 = c : b$, so that E_3F_2 is parallel to BC. The analogous property holds for E_1D_2 and D_3F_1. Moreover, $AD_2 = b\sin^2\gamma$ and $AD_3 = c\sin^2\beta$, giving

$$AD_2 : AD_3 = \sin\beta\,\sin^2\gamma : \sin\gamma\,\sin^2\beta = c : b\,.$$

AD_2D_3 is therefore similar to ABC. That is, D_2D_3 is *antiparallel* to BC (and therefore parallel to EF).

We have $AD_2 = b\sin^2\gamma = (bc/4R^2)c$ and $AD_3 = (bc/4R^2)b$, whence it follows that the scaling factor from ABC to AD_2D_3 is equal to $bc/4R^2$. But then $D_2D_3 = abc/4R^2 = [ABC]/R$, giving the elegant conclusion

$$D_2D_3 = E_3E_1 = F_1F_2\,.$$

If we walk through the six points in the order $F_1\,E_1\,D_2\,F_2\,E_3\,D_3$, we obtain a hexagon with the property that each side is parallel to the opposite one. The diagonals are antiparallel to the sides of the triangle, and are of equal length.

6

Coordinate Systems with Respect to a Triangle

6.1

In the preceding chapters, our tools have been primarily "geometric": congruence, similarity, inscribed angles, trigonometric functions. We can also take an "analytic" approach to geometry, representing points by coordinates, and lines and curves by equations. The most commonly used coordinate systems are the Cartesian and polar coordinates.

The most important figure in elementary plane geometry is the *triangle*, and much of what is treated in this book relates to it. Because of its importance, there exist coordinate systems that are associated to the triangle that is being considered. In this chapter, we discuss two of these systems.

If P is a point in the plane of triangle ABC, we will denote the distances from P to respectively BC, CA, and AB by $\bar{x}$, $\bar{y}$, and $\bar{z}$, where $\bar{x}$ is considered to be positive if P lies on the same side of BC as A, and the analogue holds

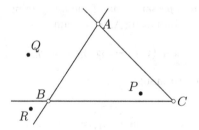

Fig. 6.1.

for $\bar{y}$ and $\bar{z}$. In Figure 6.1, $\bar{x}$, $\bar{y}$, and $\bar{z}$ are positive for P; for Q, $\bar{x}$ and $\bar{y}$ are positive, and $\bar{z}$ is negative; and for R, $\bar{y}$ is positive, and $\bar{x}$ and $\bar{z}$ negative.

It is clear that $\bar{x}$, $\bar{y}$, and $\bar{z}$ are unsuitable as coordinates as they are not linearly independent; indeed, we have

O. Bottema, *Topics in Elementary Geometry*,
DOI: 10.1007/978-0-387-78131-0_6, © Springer Science+Business Media, LLC 2008

$$a\bar{x} + b\bar{y} + c\bar{z} = 2[ABC] \,. \tag{6.1}$$

Leaving one out would break the symmetry and discriminate against one of the sides. This problem was solved in the following ingenious way: we define the *trilinear coordinates* of a point P to be any ordered set of three numbers x, y, and z for which

$$x : y : z = \bar{x} : \bar{y} : \bar{z} \,. \tag{6.2}$$

We note that this set is not unique. Indeed, for any $\lambda \neq 0$, the set $(\lambda x, \lambda y, \lambda z)$ also satisfies (6.2). Let us show that (x, y, z) determines P and is therefore well defined as a coordinate. By definition, there exists a $k \neq 0$ such that $\bar{x} = kx$ and so on. Thanks to (6.1), we have

$$k = \frac{2[ABC]}{ax + by + cz} \,,$$

which shows that unless $ax + by + cz = 0$, (x, y, z) indeed determines P. To avoid exceptions, we add the "line at infinity" with equation $ax + by + cz = 0$ to the "normal plane". "Geometric" motivations, such as the use of the expression "two parallel lines meet each other at a point at infinity", have contributed to this decision. We call the plane extended by one line the *projective* plane; it will come up again in this book.

Let us determine the equation of an arbitrary line. Because of the definition of the trilinear coordinates, the equation of a line or of a curve only makes sense geometrically if it is homogeneous in x, y, and z. Let P_1 and P_2 be two points with respective distances from BC equal to $\bar{x}_1$ and $\bar{x}_2$, and let P be a point on $P_1 P_2$ such that $P_1 P : P_1 P_2 = \lambda$. For the distance $\bar{x}$ from P to BC, we find $\bar{x} = (1 - \lambda)\bar{x}_1 + \lambda \bar{x}_2$, and in general, for P,

$$x : y : z = \left[(1 - \lambda)x_1 + \lambda x_2\right] : \left[(1 - \lambda)y_1 + \lambda y_2\right] : \left[(1 - \lambda)z_1 + \lambda z_2\right] \,, \tag{6.3}$$

which is the *parametric representation* of the line joining $P_1 = (x_1, y_1, z_1)$ and $P_2 = (x_2, y_2, z_2)$. After eliminating λ, we obtain

$$(y_1 z_2 - y_2 z_1)x + (z_1 x_2 - z_2 x_1)y + (x_1 y_2 - x_2 y_1)z = 0 \tag{6.4}$$

or, in the form of a determinant,

$$\begin{vmatrix} x & y & z \\ x_1 & y_1 & z_1 \\ x_2 & y_2 & z_2 \end{vmatrix} = 0 \,. \tag{6.5}$$

This leads to the conclusion that in trilinear coordinates, the equation of a line is linear (and homogeneous). From (6.4), we deduce that the equation is not only homogeneous in x, y, and z, but also in x_1, y_1, and z_1, and in x_2, y_2, and z_2. Conversely, a homogeneous linear equation $ux + vy + wz = 0$ represents a line: it joins the points $(0, w, -v)$ and $(-w, 0, u)$, respectively $(v, -u, 0)$.

6.2

A second coordinate system is that of the *barycentric* coordinates (barycenter is another word for center of mass). If ABC is the triangle and P is a point, then $\bar{X}$, $\bar{Y}$, and $\bar{Z}$ are defined respectively as the areas of the triangles PBC, PCA, and PAB, where PBC is considered to be positive if the sense of rotation induced by the order PBC is the same as that induced by ABC, and negative if it is the opposite. In Fig. 6.1, PBC, PCA, and PAB are positive, QBC and QCA positive, QAB negative, RBC and RAB negative, and RCA positive. There is a simple relation between trilinear and barycentric coordinates:

$$\bar{X} = a\bar{x}, \ \bar{Y} = b\bar{y}, \ \bar{Z} = c\bar{z} \ . \tag{6.6}$$

If we introduce *homogeneous barycentric coordinates* X, Y, and Z by setting, as in (6.2),

$$X : Y : Z = \bar{X} : \bar{Y} : \bar{Z} \ , \tag{6.7}$$

then

$$X : Y : Z = ax : by : cz \ . \tag{6.8}$$

In the coordinates X, Y, and Z, the line at infinity has equation $X+Y+Z = 0$. In these coordinates, a line has a homogeneous linear equation, and every homogeneous linear equation represents a line.

6.3

Let us determine the trilinear coordinates of a number of special points, and the equations of a number of lines. The barycentric coordinates immediately follow using (6.8).

The vertices A, B, and C are $(1,0,0)$, $(0,1,0)$, and $(0,0,1)$, respectively, and the sides are given by $x = 0$, $y = 0$, and $z = 0$. The center I of the incircle is $(1,1,1)$, and the centers of the excircles are $(-1,1,1)$, $(1,-1,1)$, and $(1,1,-1)$.

The centroid G has barycentric coordinates $(1,1,1)$, whence the name. The trilinear coordinates of G are (bc, ca, ab). The midpoints M_1, M_2, and M_3 of the sides are $(0,c,b)$, $(c,0,a)$, and $(b,a,0)$. The medians are given by $by - cz = 0$, $cz - ax = 0$, and $ax - by = 0$. These results contain an analytic proof of the concurrence of the three medians.

For the center O of the circumcircle (Fig. 6.2), we have $\angle BOC = 2\alpha$, $\angle BOM_1 = \alpha$, and $BO = R$, hence $OM_1 = R\cos\alpha$. In trilinear coordinates, this gives $O = (\cos\alpha, \cos\beta, \cos\gamma)$.

In Fig. 6.3, we see the altitudes AH_1 and BH_2 that meet at H. We have $BH_1 = c\cos\beta$, and since $\angle BHH_1 = \gamma$, we find that for H, $\bar{x} = c\cos\beta \cot\gamma = 2R\cos\beta \cos\gamma$. It follows that in trilinear coordinates, $H = (\cos\beta\cos\gamma, \cos\gamma\cos\alpha, \cos\alpha\cos\beta)$.

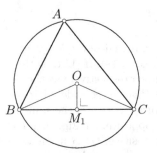

Fig. 6.2.

We have thus determined the trilinear coordinates of the three classical special points G, O, and H. From the results obtained above, we deduce the equation

$$\sum (b^2 - c^2)(b^2 + c^2 - a^2)ax = 0 \qquad (6.9)$$

for the line OH, of which we can show by simple expansion that it passes through the centroid G.

This concludes the analytic proof of the collinearity of G, O, and H.

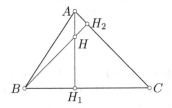

Fig. 6.3.

7

The Area of a Triangle as a Function of the Barycentric Coordinates of Its Vertices

7.1

Let ABC be a triangle taken as reference, and $P_1P_2P_3$ a triangle whose vertices have homogeneous barycentric coordinates $P_i = (X_i, Y_i, Z_i)$ for $i = 1$, 2, 3. We are going to prove the following equality:

$$[P_1P_2P_3] = \frac{\begin{vmatrix} X_1 & Y_1 & Z_1 \\ X_2 & Y_2 & Z_2 \\ X_3 & Y_3 & Z_3 \end{vmatrix}}{\prod(X_i + Y_i + Z_i)}[ABC]. \tag{7.1}$$

For simplicity we take $[ABC] = 1$. The normalized coordinates (*areal coordinates*) of a point P are given by

$$\bar{X} = [PBC], \ \bar{Y} = [PCA], \ \bar{Z} = [PAB],$$

which satisfy

$$\bar{X} + \bar{Y} + \bar{Z} = 1,$$

and in terms of which (7.1) becomes

$$[P_1P_2P_3] = \begin{vmatrix} \bar{X}_1 & \bar{Y}_1 & \bar{Z}_1 \\ \bar{X}_2 & \bar{Y}_2 & \bar{Z}_2 \\ \bar{X}_3 & \bar{Y}_3 & \bar{Z}_3 \end{vmatrix}. \tag{7.2}$$

Let $P_1 = (\bar{X}_1, \bar{Y}_1, \bar{Z}_1)$ be an arbitrary point and $S = (\bar{X}_0, \bar{Y}_0, 0)$ a point on AB (Fig. 7.1). As $\bar{X}_0 + \bar{Y}_0 = 1$, we have $AS : AB = \bar{Y}_0$, and therefore

$$[P_1AS] = \bar{Y}_0 \bar{Z}_1. \tag{7.3}$$

Next, we add a second arbitrary point P_2 such that P_1P_2 is not parallel to AB, so that $\bar{Z}_1 \neq \bar{Z}_2$. Let S be the intersection point of P_1P_2 and AB. Then (Fig. 7.1)

O. Bottema, *Topics in Elementary Geometry*,
DOI: 10.1007/978-0-387-78131-0_7, © Springer Science+Business Media, LLC 2008

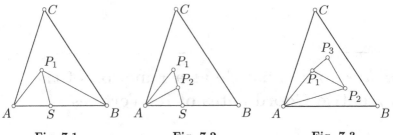

Fig. 7.1.   Fig. 7.2.   Fig. 7.3.

$$S = (\bar{Z}_2\bar{X}_1 - \bar{Z}_1\bar{X}_2, \bar{Z}_2\bar{Y}_1 - \bar{Z}_1\bar{Y}_2, 0) .$$

As

$$\bar{Z}_2\bar{X}_1 - \bar{Z}_1\bar{X}_2 + \bar{Z}_2\bar{Y}_1 - \bar{Z}_1\bar{Y}_2 = \bar{Z}_2(1 - \bar{Z}_1) - \bar{Z}_1(1 - \bar{Z}_2) = \bar{Z}_2 - \bar{Z}_1 ,$$

we have $S = (\bar{X}_0, \bar{Y}_0, 0)$ with

$$\bar{Y}_0 = \frac{\bar{Z}_2\bar{Y}_1 - \bar{Z}_1\bar{Y}_2}{\bar{Z}_2 - \bar{Z}_1} .$$

It follows from (7.3) that

$$[P_1AS] = \frac{(\bar{Z}_2\bar{Y}_1 - \bar{Z}_1\bar{Y}_2)\bar{Z}_1}{\bar{Z}_2 - \bar{Z}_1} \text{ and } [P_2AS] = \frac{(\bar{Z}_2\bar{Y}_1 - \bar{Z}_1\bar{Y}_2)\bar{Z}_2}{\bar{Z}_2 - \bar{Z}_1} ,$$

and therefore

$$[AP_1P_2] = \bar{Y}_1\bar{Z}_2 - \bar{Y}_2\bar{Z}_1 . \tag{7.4}$$

For the third and last step, we add a third arbitrary point P_3 such that P_1P_3 is not parallel to AC and P_2P_3 is not parallel to BC.

Let P_1, P_2, and P_3 be arbitrary (Fig. 7.3). Using (7.4), we have

$$[P_1P_2P_3] = [AP_2P_3] + [AP_3P_1] + [AP_1P_2]$$
$$= \bar{Y}_2\bar{Z}_3 - \bar{Y}_3\bar{Z}_2 + \bar{Y}_3\bar{Z}_1 - \bar{Y}_1\bar{Z}_3 + \bar{Y}_1\bar{Z}_2 - \bar{Y}_2\bar{Z}_1$$
$$= \begin{vmatrix} 1 & \bar{Y}_1 & \bar{Z}_1 \\ 1 & \bar{Y}_2 & \bar{Z}_2 \\ 1 & \bar{Y}_3 & \bar{Z}_3 \end{vmatrix} = \begin{vmatrix} \bar{X}_1 + \bar{Y}_1 + \bar{Z}_1 & \bar{Y}_1 & \bar{Z}_1 \\ \bar{X}_2 + \bar{Y}_2 + \bar{Z}_2 & \bar{Y}_2 & \bar{Z}_2 \\ \bar{X}_3 + \bar{Y}_3 + \bar{Z}_3 & \bar{Y}_3 & \bar{Z}_3 \end{vmatrix} = \begin{vmatrix} \bar{X}_1 & \bar{Y}_1 & \bar{Z}_1 \\ \bar{X}_2 & \bar{Y}_2 & \bar{Z}_2 \\ \bar{X}_3 & \bar{Y}_3 & \bar{Z}_3 \end{vmatrix} ,$$

which proves (7.2), and therefore also (7.1).

Let us give two applications of (7.1).

7.2

Let I_0 be the center of the incircle of triangle ABC, and let I_1, I_2, and I_3 be the centers of the excircles. Let r_0, r_1, r_2, and r_3 be the radii of these circles, and let R be the radius of the circumcircle. Then in areal coordinates, we have

$$I_0 = (ar_0, br_0, cr_0), \qquad I_1 = (-ar_1, br_1, cr_1),$$
$$I_2 = (ar_2, -br_2, cr_2), \qquad I_3 = (ar_3, br_3, -cr_3).$$

As

$$\begin{vmatrix} -ar_1 & br_1 & cr_1 \\ ar_2 & -br_2 & cr_2 \\ ar_3 & br_3 & -cr_3 \end{vmatrix} = 4abcr_1r_2r_3,$$

we deduce from (7.1) that

$$[I_1 I_2 I_3] = \frac{abc[ABC]}{2(s-a)(s-b)(s-c)} = \frac{abc\,s}{2[ABC]} = 2Rs, \qquad (7.5)$$

an elegant formula for the area of the triangle that has the centers of the excircles as vertices.

Analogously, we find

$$[I_0 I_2 I_3] = 2R(s-a), \ [I_0 I_3 I_1] = 2R(s-b), \text{ and } [I_0 I_1 I_2] = 2R(s-c).$$

7.3

Our second example concerns the area of triangle GOI, whose vertices are the triangle centroid, the center of the circumcircle, and the center of the incircle of triangle ABC.

In areal coordinates, we have

$$G = (1, 1, 1), O = (a \cos \alpha, b \cos \beta, c \cos \gamma), \text{ and } I = (a, b, c), \qquad (7.6)$$

and the determinant D in (7.1) becomes

$$\begin{aligned} D &= \begin{vmatrix} 1 & 1 & 1 \\ a \cos \alpha & b \cos \beta & c \cos \gamma \\ a & b & c \end{vmatrix} \\ &= \sum bc(\cos \beta - \cos \gamma) \\ &= \frac{1}{2abc} \sum [(a^2 - b^2 + c^2)b^2c - (a^2 + b^2 - c^2)bc^2]. \end{aligned}$$

After expansion, the sum contains 18 terms, of which 6 cancel against each other. Regrouping the remaining terms gives $\sum[a^4(c-b) + a^3(c^2 - b^2)]$, which finally reduces to

$$D = \frac{1}{2abc}[(a+b+c)^2(b-c)(c-a)(a-b)].$$

This confirms that $D = 0$ for an isosceles triangle ABC.

For the denominator of (7.1), we have

$$\prod(X_i + Y_i + G_i) = 3(a + b + c)\sum a\cos\alpha .$$

As

$$\sum a\cos\alpha = \frac{1}{2abc}\sum a^2(-a^2 + b^2 + c^2) = \frac{1}{2abc}16[ABC]^2 ,$$

our final result is

$$[GOI] = \frac{(a + b + c)(b - c)(c - a)(a - b)}{48[ABC]} . \tag{7.7}$$

This gives the impression that $[GOI]$ depends on the orientation of ABC, which seems contradictory. However, if we replace ABC by ACB, that is, exchange b and c, the area $[GOI]$ remains unchanged.

8

The Distances from a Point to the Vertices of a Triangle

8.1

In Chapter 6, the distances from a point to the *sides* of a triangle were used to define the simple and useful system of trilinear coordinates. Two things made this possible: (1) there is an obvious way to attribute a sign, or direction, to distances; and (2) any two of the distances from a point to the sides of a triangle determine the third one unambiguously.

The distances ρ_1, ρ_2, and ρ_3 from a point P to the *vertices* of ABC do not have these properties. If PA and PB are given, then, as an intersection point of two circles, P can be either real or imaginary, and in the first case, we find two values for PC. There will be a relation between $PA = d_1$, $PB = d_2$, and $PC = d_3$ and the sides a, b, and c, which we expect to be quadratic. There are different methods for determining this relation. We choose the following one (Fig. 8.1), where P is chosen inside ABC and $\angle BPC = \varphi_1$, $\angle CPA = \varphi_2$, and

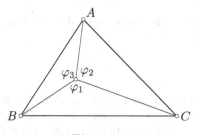

Fig. 8.1.

$\angle APB = \varphi_3$. We have $\varphi_1 + \varphi_2 + \varphi_3 = 2\pi$ and therefore $\cos(\varphi_1 + \varphi_2 + \varphi_3) = 1$, whence it follows from trigonometric addition formulas that

$$2 \cos\varphi_1 \, \cos\varphi_2 \, \cos\varphi_3 - \cos^2\varphi_1 - \cos^2\varphi_2 - \cos^2\varphi_3 + 1 = 0 \, . \qquad (8.1)$$

O. Bottema, *Topics in Elementary Geometry*,
DOI: 10.1007/978-0-387-78131-0_8, © Springer Science+Business Media, LLC 2008

Since

$$\cos\varphi_1 = \frac{-a^2 + d_2^2 + d_3^2}{2d_2d_3}$$

and so on, we find

$$(-a^2 + d_2^2 + d_3^2)(-b^2 + d_3^2 + d_1^2)(-c^2 + d_1^2 + d_2^2)$$
$$- (-a^2 + d_2^2 + d_3^2)^2 d_1^2 - (-b^2 + d_3^2 + d_1^2)^2 d_2^2 - (-c^2 + d_1^2 + d_2^2)^2 d_3^2$$
$$+ 4d_1^2 d_2^2 d_3^2 = 0 \,,$$
(8.2)

a relation that remains valid when P lies outside the triangle. Relation (8.2) is of degree *three* in the d_i^2, which is higher than we expected. However, it turns out that the terms of degree three in the d_i^2, such as $d_1^2 d_2^2 d_3^2$ and $d_1^4 d_2^2$, cancel out. This also follows from the fact that the left-hand side of (8.2) is identically zero for $a = b = c = 0$.

After some computation, (8.2) can finally be written as

$$a^2 d_1^2(-d_1^2 + d_2^2 + d_3^2 - a^2 + b^2 + c^2)$$
$$+ b^2 d_2^2(d_1^2 - d_2^2 + d_3^2 + a^2 - b^2 + c^2) + c^2 d_3^2(d_1^2 + d_2^2 - d_3^2 + a^2 + b^2 - c^2)$$
$$- (a^2 d_2^2 d_3^2 + b^2 d_3^2 d_1^2 + c^2 d_1^2 d_2^2 + a^2 b^2 c^2) = 0 \,.$$
(8.3)

This complicated formula with 22 terms is in great contrast to the relation $ad_1 + bd_2 + cd_3 = 2[ABC]$ we found for the distances to the sides. Let us give a couple of examples.

In an equilateral triangle with side a, we have

$$\sum d_i^4 - \sum d_j^2 d_k^2 - a^2 \sum d_i^2 + a^4 = 0 \,.$$

If in (8.3) we let $d_1 = d_2 = d_3 = d$, we obtain $d^2 = (a^2 b^2 c^2 / 16[ABC]^2) R^2$.

8.2

Let us describe a problem related to (8.3), which we will come back to in Chapter 23. Consider a triangle ABC and three positive numbers p, q, and r. We would like to know whether there exist points T for which

$$TA : TB : TC = p : q : r \,.$$
(8.4)

We can show as follows that the locus of the points T for which $TB : TC = q : r$ is a circle, called an APOLLONIUS *circle*. If T is one of the points that we are looking for, the interior and exterior angle bisectors of the vertex angle T meet the base BC at the points S_1 and S_2 for which $BS : SC = q : r$, which implies that S_1 and S_2 are fixed points. As TS_1 and TS_2 are perpendicular to each other, the desired locus is the circle with diameter S_1S_2. It follows that the points that satisfy (8.4) are the intersection points of two, and therefore three, APOLLONIUS circles. We can consequently expect two, one, or zero real solutions. To determine the criteria for this, we return to (8.3).

Let $TA = d_1 = pt$ and consequently $d_2 = qt$ and $d_3 = rt$. Substitution in (8.3) gives the following quadratic equation in t^2:

$$Ut^4 + Vt^2 + W = 0 , \tag{8.5}$$

where

$$
\begin{aligned}
U &= -a^2p^4 + a^2p^2q^2 + a^2p^2r^2 - b^2q^4 + b^2q^2r^2 + b^2q^2p^2 \\
&\quad - c^2r^4 + c^2r^2p^2 + c^2r^2q^2 - a^2q^2r^2 - b^2r^2p^2 - c^2p^2q^2 \\
&= -a^2p^4 - b^2q^4 - c^2r^4 \\
&\quad + 2bcq^2r^2 \cos\alpha + 2car^2p^2 \cos\beta + 2abp^2q^2 \cos\gamma , \\
V &= 2abc(ap^2 \cos\alpha + bq^2 \cos\beta + cr^2 \cos\gamma) , \\
W &= -a^2b^2c^2 .
\end{aligned}
\tag{8.6}
$$

Let D' be the discriminant $D = V^2 - 4UW$ divided by $4a^2b^2c^2$. We find

$$
\begin{aligned}
D' &= a^2p^4 \cos^2\alpha + b^2q^4 \cos^2\beta + c^2r^4 \cos^2\gamma \\
&\quad + 2bcq^2r^2 \cos\beta \cos\gamma + 2car^2p^2 \cos\gamma \cos\alpha + 2abp^2q^2 \cos\alpha \cos\beta \\
&\quad - a^2p^4 - b^2q^4 - c^2r^4 \\
&\quad + 2bcq^2r^2 \cos\alpha + 2car^2p^2 \cos\beta + 2abp^2q^2 \cos\gamma .
\end{aligned}
$$

As $1 - \cos^2\alpha = \sin^2\alpha$ and so on, and $\cos\alpha + \cos\beta \cos\gamma = \sin\beta \sin\gamma$ and so on, this gives

$$
\begin{aligned}
D' &= -a^2p^4 \sin^2\alpha - b^2q^4 \sin^2\beta - c^2r^4 \sin^2\gamma \\
&\quad + 2bcq^2r^2 \sin\beta \sin\gamma + 2car^2p^2 \sin\gamma \sin\alpha + 2abp^2q^2 \sin\alpha \sin\beta ,
\end{aligned}
$$

or, finally, as $\sin\alpha = a/2R$ and so on:

$$D' = \frac{1}{4R^2}(-a^4p^4 - b^4q^4 - c^4r^4 + 2b^2c^2q^2r^2 + 2c^2a^2r^2p^2 + 2a^2b^2p^2q^2) . \tag{8.7}$$

If we recall that three positive numbers u, v, and w are the sides of a triangle only if $-u^4 - v^4 - w^4 + 2v^2w^2 + 2w^2u^2 + 2u^2v^2 \geq 0$, where equality holds for a degenerate triangle, it follows that the roots t^2 of (8.5) are real only if there exists a triangle with sides ap, bq, and cr.

Let us continue the discussion of (8.5) by determining the signs of U and V. The first equality in (8.6) implies that

$$
\begin{aligned}
-2U &= (-a^2 + b^2 + c^2)(q^2 - r^2)^2 + (a^2 - b^2 + c^2)(r^2 - p^2)^2 \\
&\quad + (a^2 + b^2 - c^2)(p^2 - q^2)^2 .
\end{aligned}
\tag{8.8}
$$

If triangle ABC is not obtuse, U is negative. If ABC is obtuse, let us say in C, then $c^2 > a^2 + b^2$ and only the third term of (8.8) is negative. Let us set $q^2 - r^2 = K$ and $r^2 - p^2 = L$. Then $(p^2 - q^2)^2 = (K + L)^2$ and

$$-2U = (-a^2 + b^2 + c^2)K^2 + (a^2 - b^2 + c^2)L^2 + (a^2 + b^2 - c^2)(K + L)^2$$
$$= 2b^2 K^2 + 2(a^2 + b^2 - c^2)KL + 2a^2 L^2 ,$$

which, as $c < a + b$, is greater than

$$2b^2 K^2 - 4abKL + 2a^2 L^2 = 2(bK - aL)^2 .$$

U is therefore negative, or zero if $p = q = r$.

As far as V is concerned, its coefficient is clearly positive for an angle that is not obtuse. If C is obtuse, the third term is negative. If the condition under which the roots t^2 are real is satisfied, then $rc < pa + qb$ and

$$V > p^2 a^2 (-a^2 + b^2 + c^2) + q^2 b^2 (a^2 - b^2 + c^2) + (pa + qb)^2 (a^2 + b^2 - c^2)$$
$$= 2p^2 a^2 b^2 + 2pqab(a^2 + b^2 - c^2) + 2q^2 a^2 b^2 ,$$

and therefore, as $c < a + b$,

$$V \geq 2p^2 a^2 b^2 - 4pqa^2 b^2 + 2q^2 a^2 b^2 = 2a^2 b^2 (p - q)^2 \geq 0 .$$

It follows from $U < 0$, $V > 0$, $W < 0$, and $D > 0$ that both roots t^2 of (8.5) are positive, giving two positive roots for t. We have shown, using a difficult but elementary method requiring much computation, that *there exist two real points T_i satisfying (8.4) if and only if ap, bq, and cr satisfy the triangle inequalities.*

8.3

By what we have just seen, the distances r_1, r_2, and r_3 from a point P to the vertices of a triangle ABC are unsuitable for use as a coordinate system. Their ratios cannot be used either: they correspond to two, one, or zero real points. They are no true rival for the trilinear or barycentric coordinates. However, the elaborate computations of this chapter have not been completely in vain. There are questions of geometric type for which they can be used. Let us give two examples.

A first example: do there exist, in the plane of triangle ABC, points for which the distances to the vertices are *inversely proportional* to the opposite sides?

It immediately follows from $p = a^{-1}$, $q = b^{-1}$, and $r = c^{-1}$ with $ap = bq = cr$ that for every triangle there exist two real points I_1 and I_2 with the desired property. They are called *isodynamic points* of the triangle.

A second example: are there points for which the distances are *proportional* to the opposite sides? The answer is that there are two such real points Q_1 and Q_2 only if a^2, b^2, and c^2 satisfy the triangle inequalities, that is, if the triangle is *acute*.

A completely different application is the following. The condition under which two circles with radii R_1 and R_2 and distance d between their centers meet perpendicularly, is that $R_1^2 + R_2^2 = d^2$. If $R_2 = 0$, the relation becomes $d^2 = R_1^2$. That is, the center of a point circle which meets a true circle perpendicularly lies on that circle's circumference.

We apply this to the point circles $\rho_1^2 = 0$, $\rho_2^2 = 0$, and $\rho_3^2 = 0$ that have respective centers A, B, and C, and therefore all meet the circumcircle Ω with center O perpendicularly. It follows that every circle of the family determined by the equation

$$\lambda_1 \rho_1^2 + \lambda_2 \rho_2^2 + \lambda_3 \rho_3^2 = 0 \tag{8.9}$$

also meets Ω perpendicularly. This family also contains lines, circles with infinite radius. Such a "circle" must pass through O, for which $\rho_1 = \rho_2 = \rho_3$. It follows that the lines passing through O are given by (8.9) with the condition

$$\lambda_1 + \lambda_2 + \lambda_3 = 0 \,. \tag{8.10}$$

The median through A satisfies $m_a^2 = \frac{1}{2}b^2 + \frac{1}{2}c^2 - \frac{1}{4}a^2$. As $GA = \frac{2}{3}m_a$, we find the following ρ-coordinates for G:

$$\rho_1^2 : \rho_2^2 : \rho_3^2 = (2b^2 + 2c^2 - a^2) : (2c^2 + 2a^2 - b^2) : (2a^2 + 2b^2 - c^2) \,. \tag{8.11}$$

The line given by (8.9) passes through G if the following relation holds in addition to (8.10):

$$(2b^2 + 2c^2 - a^2)\lambda_1 + (2c^2 + 2a^2 - b^2)\lambda_2 + (2a^2 + 2b^2 - c^2)\lambda_3 = 0 \,. \tag{8.12}$$

The ratios between λ_1, λ_2, and λ_3 follow from (8.10) and (8.12). We conclude that *the equation of the* EULER *line in ρ-coordinates is*

$$(b^2 - c^2)\rho_1^2 + (c^2 - a^2)\rho_2^2 + (a^2 - b^2)\rho_3^2 = 0 \,. \tag{8.13}$$

Let us give one application: $\rho_1^2 : \rho_2^2 : \rho_3^2 = a^2 : b^2 : c^2$ satisfies (8.13).

It follows that *the points denoted above by Q_1 and Q_2, which are real in any acute triangle, lie on the* EULER *line*. From (8.13) it follows, as expected, that the EULER line is not well defined in an equilateral triangle.

9

The Simson Line

9.1

Let l be a line (Fig. 9.1) that meets the sides of triangle ABC in such a

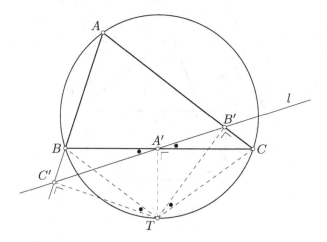

Fig. 9.1.

way that the perpendiculars at the intersections points A', B', and C' concur at a point T. If we also draw TB and TC, then $TC'BA'$ and $TCB'A'$ are cyclic quadrilaterals, which implies that the angles $C'TB$ and CTB' are equal, respectively, to the angles $BA'C'$ and $B'A'C$. In particular, they are equal to each other. The same holds for angles $C'TB'$ and BTC. The first of these is the supplement of angle BAC, as $C'TB'A$ is a cyclic quadrilateral. It follows that BTC and BAC are also supplementary angles, so that T lies on the excircle of the triangle.

O. Bottema, *Topics in Elementary Geometry*,
DOI: 10.1007/978-0-387-78131-0_9, © Springer Science+Business Media, LLC 2008

Repeating the arguments in another order, we can convince ourselves that the converse of this statement is also correct. Hence, *the circumcircle of the triangle is the locus of the points whose projections on the sides are collinear.* This way, a line l_T is associated to each point T of the circle. It is called the SIMSON *line* of T (1798), or sometimes the WALLACE-SIMSON *line* of T [Wal], [Joh3]. We note that this theorem is a special case of Section 3.7.

9.2

Let (Fig. 9.2) H be the orthocenter of triangle ABC and let H'_a be the second intersection point of altitude AA' with the circumcircle. Then $HA' = A'H'_a$, as follows, for example, from the properties of the nine-point circle. We can

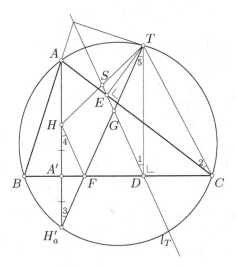

Fig. 9.2.

now conclude that the following equalities hold, where the angles are indicated with φ_i and refer to the figure: $\varphi_1 = \varphi_2$, $\varphi_2 = \varphi_3$, $\varphi_3 = \varphi_4$, and $\varphi_3 = \varphi_5$. It follows from $\varphi_1 = \varphi_5$ that $TG = DG$, so in the right triangle TDF, we have $TG = GF$. The equality $\varphi_1 = \varphi_4$ implies that l_T and FH are parallel, whence we also have $TS = SH$, or, in words: *the line segment joining T and the orthocenter H is bisected by l_T.* Of course, the midpoint S lies on the nine-point circle n. Moreover, the same figure shows that φ_1 is the angle formed by l_T and the perpendicular from T on BC, and $\varphi_2 \ (= \varphi_1)$ is half of arc AT. Therefore, if we have *two* points T_1 and T_2 on the circumcircle, the angle between l_{T_1} and l_{T_2} is half of arc T_1T_2. In particular, *diametrically opposite points* on a circle have perpendicular SIMSON lines. However, if T_1 and T_2 are diametrically opposite, S_1 and S_2 are diametrically opposite on n. It follows that the intersection point of two SIMSON lines lies on n. It is moreover worth noting that if T walks around the whole circle, the line l_T ends up in its initial position, as expected, but has only rotated over a straight angle.

Let us consider a number of special cases. First of all, we immediately see that the line associated to a vertex is the *altitude* from that vertex. Just as evident is the fact that the line associated to a point diametrically opposite to a vertex is the opposite side.

If we consider the point H_a', the associated line goes through A' and its direction is that of the line joining the projections of H on AB and AC. The SIMSON lines of H_a', H_b', and H_c' therefore go through the vertices of the pedal triangle $A'B'C'$ and are each parallel to the corresponding opposite side. Finally, if K is one of the midpoints of arc BC, the line l_K goes through the midpoint of side BC and is perpendicular to AK, which, after all, is either an interior or an exterior bisector of angle A.

10

Morley's Triangle

10.1

In 1904, F. MORLEY stated the following surprising theorem: if we draw the *angle trisectors* in a triangle ABC (that is, the lines that divide the angles in three equal parts) and intersect them with each other as shown in Figure 10.1, triangle DEF is equilateral. From approximately 1920 on, this remarkable

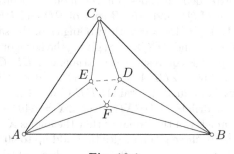

Fig. 10.1.

theorem has attracted unusual amounts of attention. Many proofs have been given, and the theorem has been extended in different directions [OB]. Maybe someday a collection of proofs will appear, as has already happened for the Pythagorean theorem.

10.2

We can give a direct proof by expressing the lengths EF, FD, and DE in the elements of the given triangle, using trigonometry. In triangle BDC, we have $\angle BDC = 120° + A/3$, so that the law of sines gives

O. Bottema, *Topics in Elementary Geometry*,
DOI: 10.1007/978-0-387-78131-0_10, © Springer Science+Business Media, LLC 2008

$$BD = \frac{a\sin(\frac{1}{3}C)}{\sin(120° + \frac{1}{3}A)} = \frac{2R\sin A\sin(\frac{1}{3}C)}{\sin(120° + \frac{1}{3}A)},$$

or, using the relation $4\sin(x/3)\sin((x + 180°)/3)\sin((x + 360°)/3) = \sin x$,

$$BD = 8R\sin\tfrac{1}{3}A\sin\tfrac{1}{3}(180° + A)\sin\tfrac{1}{3}C.$$

Cyclic permutations lead to expressions for CD, AE and CE, and AF and BF. Using the law of cosines, we can subsequently, for example, determine side EF of triangle AEF. By its symmetry, the result

$$EF = 8R\sin\tfrac{1}{3}A\,\sin\tfrac{1}{3}B\,\sin\tfrac{1}{3}C$$

proves that DEF is equilateral.

10.3

There also exist several more geometric proofs. The following one is also quite short. It uses the principle of *inversion*, which comes up from time to time in plane geometry. The opinions on the proof's merits will probably diverge.

Let DEF (Fig. 10.2) be an equilateral triangle. Let us choose points P, Q, and R on the extended altitudes, for the moment arbitrarily. Let A be the intersection point of QF and RE, B that of RD and PF, and C that of PE and QD. We flank $\angle EAF = \alpha$ with two angles of the same size, and do likewise with $\angle FBD = \beta$ and $\angle DCE = \gamma$, giving the hexagon $AZBXCY$. Let x, y, and z be the top angles of, respectively, triangles PEF, QFD, and RDE, then $\angle BDC = y + z + 60°$, $\angle CEA = z + x + 60°$, and $\angle AFB = x + y + 60°$. Moreover, it follows from quadrilateral $AFDE$ that $y + z = \alpha + 60°$, and so on. Consequently, quadrilateral $DBXC$ gives $\angle X = 240° - (\alpha + \beta + \gamma)$. It follows from the symmetry of this equality that the angles X, Y, and Z are all equal. If α, β, and γ are three given angles that add up to $60°$, then by taking $x = 60° - \alpha$, $y = 60° - \beta$, and $z = 60° - \gamma$, we obtain straight angles at X, Y, and Z. Consequently, we have a triangle with prescribed angles 3α, 3β, and 3γ built around triangle DEF, in such a way that DEF is its MORLEY triangle.

10.4

Of the many extensions of the theorem, let us just mention the one where, in addition to interior angle trisectors, we also admit exterior angle trisectors. The latter, after all, should be seen as being completely equivalent to the first (VAN IJZEREN, 1937 [IJz]; SCHUH, 1939 [Schu]).

An added difficulty in this extension consists of finding the combinations that we must choose to once more obtain an equilateral triangle formed by

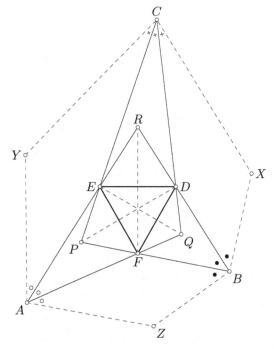

Fig. 10.2.

intersection points. Compare, on a smaller scale, the question of finding combinations with concurrent bisectors. It turns out that there are 18 MORLEY triangles, which together have 27 vertices and whose sides lie along 9 different lines, which, moreover, are parallel by threes.

Inequalities in a Triangle

11.1

There are more constructions that when applied to an arbitrary triangle result in an equilateral one. If we draw equilateral triangles *outward* on the sides of triangle ABC (Fig. 11.1), and let A', B', and C' denote their centers, then

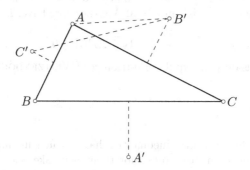

Fig. 11.1.

$$AB' = \tfrac{1}{3}b\sqrt{3}, \quad AC' = \tfrac{1}{3}c\sqrt{3}, \text{ and } \angle B'AC' = 60° + A.$$

In triangle $B'AC'$, we find:

$$(B'C')^2 = \tfrac{1}{6}(a^2 + b^2 + c^2 + 4\sqrt{3}\,[ABC]) ,$$

where $[ABC]$ denotes the area of triangle ABC. It follows from the symmetry of the expression that $A'B'C'$ is an *equilateral triangle*.

We can easily convince ourselves that the theorem *does not* hold, for example, for the centers of the squares based on the sides. The points A', B', and C' have more interesting properties; in particular, AA', BB', and CC'

O. Bottema, *Topics in Elementary Geometry*,
DOI: 10.1007/978-0-387-78131-0_11, © Springer Science+Business Media, LLC 2008

are concurrent. This property is a special case of a more general theorem. It holds whenever A', B', and C' are the tops of similar isosceles triangles drawn outward on, respectively, BC, CA, and AB. If φ is the base angle, then

$$u_1 = \frac{c \sin(B + \varphi)}{b \sin(C + \varphi)}$$

and so on. The proof now follows from CEVA's theorem.

We can also draw the three equilateral triangles *inward*. If A'', B'', and C'' are their centers, we find, completely analogously,

$$(B''C'')^2 = \frac{1}{6}(a^2 + b^2 + c^2 - 4\sqrt{3}\,[ABC])\,,$$

whence it follows that $A''B''C''$ is equilateral.

11.2

This last equality implies an inequality that holds for the elements of any triangle. Indeed, the expression for $(B''C'')^2$ cannot be negative. It can only be zero if the points A'', B'', and C'' coincide and that seems only possible if the original triangle ABC is equilateral. We therefore have: *in every triangle*,

$$a^2 + b^2 + c^2 - 4\sqrt{3}\,[ABC] \geq 0\,, \tag{11.1}$$

and equality holds only for equilateral triangles (WEITZENBÖCK, 1919 [Wei]).

11.3

During the past half century, inequalities have gained in importance. This is due to the interest in inequalities for their own sake and to a shift from the study of specific cases to more general inequalities involving functions. Another reason is the quest for a greater accuracy in modern analysis in which there are strict rules that require a discussion of the maximal deviation. Finally, inequalities have become more important because certain problems do not yet admit an exact answer, while for some of these, the search for the exact answer is not even truly interesting. Geometric figures have gradually given rise to a great number of inequalities of which, in 1969, an extensive collection was published as a book [Bot2].

11.4

The inequality mentioned above arose directly from the interpretation of a particular geometric formula. In an analogous manner, other inequalities for the elements of a triangle can be deduced. For example, the distance HO between the orthocenter and the center of the circumcircle satisfies

$$HO^2 = R^2(1 - 8\cos A \cos B \cos C) ,$$

from which the inequality

$$\cos A \cos B \cos C \leq \frac{1}{8} \qquad (11.2)$$

follows. Equality holds only for the equilateral triangle. In Chapter 14, we will prove the relation $IO^2 = R^2 - 2Rr$ for the distance between the centers of the incircle and circumcircle, of which the inequality

$$R \geq 2r \qquad (11.3)$$

is a direct consequence.

11.5

As $R = abc/(4[ABC])$ and $r = 2[ABC]/(a+b+c)$, we can also write (11.3) in the form $abc(a + b + c) - 16[ABC]^2 \geq 0$. A known theorem from algebra says that the geometric mean of a set of positive numbers is at most equal to their arithmetic mean, and that equality occurs only if all numbers are equal. Therefore,

$$abc \leq \frac{(a+b+c)^3}{27} ,$$

from which it follows that $(a + b + c)^4 - 16 \times 27 \times [ABC]^2 \geq 0$, that is,

$$(a + b + c)^2 - 12\sqrt{3}\,[ABC] \geq 0. \qquad (11.4)$$

Equality holds only for equilateral triangles. Inequality (11.4) gives a relation between the *perimeter* p and the *area* $[F]$ of certain figures, in this case triangles. A geometric inequality of this type is called an *isoperimetric inequality*. Their general appearance is

$$p^2 - c[F] \geq 0 ,$$

where c denotes a constant that depends on the type of figure that is being considered.

It immediately follows from the easily verified identity

$$\begin{aligned} 3(a^2 + b^2 + c^2) &= (a + b + c)^2 + (a - b)^2 + (b - c)^2 + (c - a)^2 \\ &= (a + b + c)^2 + Q \qquad (11.5) \\ &\geq (a + b + c)^2 \end{aligned}$$

that (11.1) results from (11.4). In this manner we can also deduce the following equation, called a *sharpening* of (11.1):

$$a^2 + b^2 + c^2 - 4\sqrt{3}\,[ABC] \geq \tfrac{1}{3}Q\,. \tag{11.6}$$

It can be shown using elementary methods that the factor $1/3$ on the right-hand side may be replaced by 1. Consequently, the following sharpening of the isoperimetric inequality for triangles can be shown in an analogous manner:

$$(a + b + c)^2 - 12\sqrt{3}\,[ABC] \geq 2Q \tag{11.7}$$

(FINSLER and HADWIGER, 1937 [FH]).

We have $r = 2[ABC]/(a + b + c)$, or by (11.4),

$$r \leq \frac{1}{6\sqrt{3}}(a + b + c)\,, \tag{11.8}$$

whence, after squaring and applying (11.5),

$$r^2 \leq \frac{1}{36}(a^2 + b^2 + c^2)\,, \tag{11.9}$$

an inequality that was found by KUBOTA (1923), though in a much more complex manner [Kub], [Ben]. It should be clear that seen purely algebraically, all of these inequalities are relations between a, b, and c that depend on fundamental inequalities, and which express the fact that

$$a,\ b,\ c,\ -a + b + c,\ a - b + c,\ \text{and}\ a + b - c$$

are all positive.

11.6

The inequalities are closely related to questions concerning *extrema*. For example, (11.3) implies that *the minimal value of the ratio $R : r$ between the radii of the circumcircle and of the incircle of a triangle is 2 and that this minimum is attained only in an equilateral triangle*. The isoperimetric inequality (11.4) gives rise to the theorem: *of all triangles with fixed perimeter, the equilateral triangle has the greatest area*, or: *of all triangles with fixed area, the equilateral triangle has the smallest perimeter*.

The Mixed Area of Two Parallel Polygons

12.1

Questions such as those that arose in Section 11.6 have always attracted much attention from geometers. Closely related to them are considerations such as the following, which were stated by MINKOWSKI (1903) [Min] for figures of very general type, and which we will try to present for the elementary case of a polygon.

Let $V_1 = A_1A_2 \ldots A_n$ and $V_2 = B_1B_2 \ldots B_n$ be two n-gons whose corresponding sides are parallel: $A_1A_2 \; // \; B_1B_2$, and so on. For simplicity, we've assumed in the figure that V_1 lies inside V_2. This fact is not necessary for

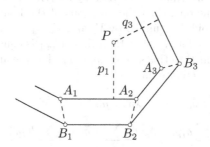

Fig. 12.1.

the theorems that follow and the proofs can remain unaltered in other situations, if we take care to attribute the correct signs to the occurring entities. Let $A_1A_2 = a_1$, $A_2A_3 = a_2, \ldots$, and $B_1B_2 = b_1$, $B_2B_3 = b_2, \ldots$. Let P be an arbitrary point, chosen inside V_1 in our figure, and let $p_1, p_2, \ldots$ be the distances from A_1A_2, $A_2A_3, \ldots$ and $q_1, q_2, \ldots$, those from B_1B_2, $B_2B_3, \ldots$. Then the areas of V_1 and V_2 satisfy $[V_1] = \sum a_ip_i/2$ and $[V_2] = \sum b_iq_i/2$, respectively. The area $[V_2]$ is obtained by adding the area of a ring of trapezoids

O. Bottema, *Topics in Elementary Geometry*,
DOI: 10.1007/978-0-387-78131-0_12, © Springer Science+Business Media, LLC 2008

such as $A_1 B_1 B_2 A_2$ to $[V_1]$. The area of this trapezoid is $(a_1 + b_1)(q_1 - p_1)/2$. We therefore have

$$[V_2] = [V_1] + \tfrac{1}{2} \sum (a_i + b_i)(q_i - p_i) \, ,$$

that is,

$$[V_2] = [V_1] + \tfrac{1}{2} \sum b_i q_i - \tfrac{1}{2} \sum a_i p_i + \tfrac{1}{2} \sum a_i q_i - \tfrac{1}{2} \sum b_i p_i \, .$$

It follows from this that $\sum a_i q_i = \sum b_i p_i$, a strange relation which, in words, reads: *if we multiply each side of V_1 by the distance to the corresponding side of V_2, then the sum of these products is the same as that which is obtained when V_1 and V_2 are interchanged.*

12.2

We take the point C_i on the segment $A_i B_i$ such that $A_i C_i : C_i B_i = \mu : \lambda$, where $\lambda + \mu = 1$ with λ and μ real, and possibly negative. The points C_i are the vertices of a polygon $V(\lambda, \mu)$ whose sides are parallel to those of V_1 and V_2. By varying λ and μ we obtain a collection of polygons parallel to one another that we call a *linear family* and for which we also use the notation $V(\lambda, \mu) = \lambda V_1 + \mu V_2$. For $\lambda = 1$, respectively $\mu = 1$, we obtain V_1, respectively V_2. In fact, these polygons do not occupy a special position in the family, but can be replaced by two arbitrary polygons of the family, which, in turn, determine the whole family. If we think of V_1 and V_2 as lying in two parallel planes instead of in the same one, then the polygons of the family are the parallel sections of the prismoid determined by V_1 and V_2.

It now immediately follows that the side $C_i C_{i+1}$ is equal to $\lambda a_i + \mu b_i$ and that the distance from P to this side is $\lambda p_i + \mu q_i$. For the area $[V(\lambda, \mu)]$ of the polygon $V(\lambda, \mu)$, we therefore find

$$
\begin{aligned}
[V(\lambda, \mu)] &= \tfrac{1}{2} \sum (\lambda a_i + \mu b_i)(\lambda p_i + \mu q_i) \\
&= \tfrac{1}{2} \lambda^2 \sum a_i p_i + \lambda \mu \left(\tfrac{1}{2} \sum a_i q_i + \tfrac{1}{2} \sum b_i p_i \right) + \tfrac{1}{2} \mu^2 \sum b_i q_i \, ,
\end{aligned}
$$

or $[V(\lambda, \mu)] = \lambda^2 [V_1] + 2\lambda \mu [V_{1\,2}] + \mu^2 [V_2]$, where

$$[V_{1\,2}] = \tfrac{1}{2} \sum a_i q_i = \tfrac{1}{2} \sum b_i p_i.$$

As $[V(\lambda, \mu)]$, $[V_1]$, and $[V_2]$ are independent of P, we conclude that the equal expressions $\sum a_i q_i / 2$ and $\sum b_i p_i / 2$ are independent of P. We call $[V_{1\,2}]$ the *mixed area* of V_1 and V_2. Clearly, the mixed area of V_1 and V_1 is equal to $[V_1]$. If, for example, we translate V_2, then $V(\lambda, \mu)$ is only translated, hence stays congruent and keeps the same area, so that we see that *the mixed area of two polygons does not change if we shift them with respect to each other.*

12.3

If V_1 and V_2 are *homothetic*, that is, if V_2 arises from V_1 through an expansion or contraction, say with center P and scaling factor k, then $b_i = ka_i$ and $q_i = kp_i$, hence $[V_2] = k^2[V_1]$ and $[V_{1\,2}] = k[V_1]$, so that $[V_{1\,2}]^2 = [V_1][V_2]$. The fundamental theorem of MINKOWSKI now says that *this equality holds only for two homothetic polygons, and that we always have* $[V_{1\,2}]^2 \geq [V_1][V_2]$.

Let us prove this property for the particular case of two trapezoids. If

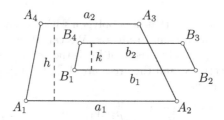

Fig. 12.2.

$A_1A_2A_3A_4$ and $B_1B_2B_3B_4$ are the two parallel trapezoids, $A_1A_2 = a_1$, $A_4A_3 = a_2$, $B_1B_2 = b_1$, $B_4B_3 = b_2$, and the heights are respectively h and k (Fig. 12.2), then

$$[V_1] = \tfrac{1}{2}(a_1 + a_2)h, \ [V_2] = \tfrac{1}{2}(b_1 + b_2)k \ .$$

Moreover,

$$[V(\lambda, \mu)] = \tfrac{1}{2}\big[(\lambda a_1 + \mu b_1) + (\lambda a_2 + \mu b_2)\big](\lambda h + \mu k)$$
$$= \lambda^2[V_1] + \tfrac{1}{2}\lambda\mu\big[(a_1 + a_2)k + (b_1 + b_2)h\big] + \mu^2[V_2] \ ,$$

whence

$$[V_{1\,2}]^2 - [V_1][V_2] = \tfrac{1}{16}\big[(a_1 + a_2)k + (b_1 + b_2)h\big]^2 - \tfrac{1}{4}(a_1 + a_2)(b_1 + b_2)hk$$
$$= \tfrac{1}{16}\big[(a_1 + a_2)k - (b_1 + b_2)h\big]^2$$
$$\geq 0 \ .$$

Equality holds only if $h : k = (a_1 + a_2) : (b_1 + b_2)$. However, $h : k = (a_1 - a_2) : (b_1 - b_2)$ always holds, so that $a_1 : a_2 = b_1 : b_2$ and $h : k = a_1 : b_1$, whence the homothety.

12.4

The expression $[V_{1\,2}]^2 - [V_1][V_2]$ is none other than the discriminant of the quadratic form $[V(\lambda, \mu)]$. MINKOWSKI's inequality therefore states that in a linear family, there exist two polygons of area zero, unless the family is made up of figures that are homothetic to one another. For our trapezoids, this can also easily be seen directly. If $\lambda : \mu = -k : h$, we obtain a trapezoid whose parallel sides lie on a single line; for $\lambda : \mu = -(b_1 + b_2) : (a_1 + a_2)$, we obtain a trapezoid whose parallel sides $A_1 A_2$ and $A_4 A_3$ have equal lengths, but opposite directions.

12.5

Let V_1 be a *convex* polygon, that is, a polygon without any reflex angles. Let C be a circle with center P and radius r. We can then draw a polygon V_2 *around* C, with sides parallel to those of V_1. If the *perimeters* of V_1 and of V_2 are p_1 and p_2, respectively, and the areas are $[V_1]$ and $[V_2]$, then $[V_2] = p_2 r / 2$ and $[V_{1\,2}] = p_1 r / 2$, so that

$$\tfrac{1}{4} p_1^2 r^2 - \tfrac{1}{2} p_2 r [V_1] \geq 0 \,,$$

or if we denote the constant $2 p_2 / r$ by c, $p_1^2 - c [V_1] \geq 0$.

Therefore, *an isoperimetric inequality holds for the convex polygons with given angles. The constant c only depends on the angles:* if these are φ_1, φ_2, ..., φ_n, then $c = 4 \sum \cot(\varphi_i / 2)$.

Equality holds only if the polygon is a tangential polygon. We conclude that *of the polygons with fixed perimeter and angles, the tangential polygon has the greatest area.*

12.6

The theory of linear families was not developed for polygons, but for arbitrary closed convex curves k_1 and k_2 (Fig. 12.3). By corresponding points A and B,

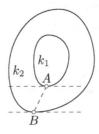

Fig. 12.3.

we mean points with parallel tangent lines. If C is the point on AB such that $AC : CB = \mu : \lambda$, then the locus of C is element $k(\lambda, \mu)$ of the family. If

$[V_1]$, $[V_2]$ and $[V(\lambda, \mu)]$ are the areas enclosed by k_1, k_2, and $k(\lambda, \mu)$, then $[V(\lambda, \mu)] = \lambda^2[V_1] + 2\lambda\mu[V_{1\,2}] + \mu^2[V_2]$, where $[V_{1\,2}]$ is once again called the mixed area of k_1 and k_2. Applying the inequality that holds for this entity to the figure of an arbitrary closed convex curve C with area $[C]$ and perimeter p and a circle with perimeter p_2, radius r, and $2p_2/r = 4\pi$ shows that the inequality

$$p^2 - 4\pi\,[C] \geq 0$$

holds for a closed convex curve. It follows from this that *of all closed convex curves with fixed perimeter, the circle has the greatest area.*

We note that the theory of linear families and the MINKOWSKI inequality have also been extended to certain non-convex figures (GEPPERT, 1937 [Gep]; BOL, 1939 [Bol1]).

13

The Isoperimetric Inequality

13.1

Let ABCD (Figs. 13.1a and 13.1b) be a quadrilateral of arbitrary form, possibly with one or two reflex angles. Let AB, BC, CD, and DA be, re-

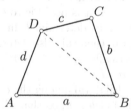

Fig. 13.1a.

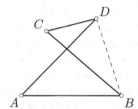

Fig. 13.1b.

spectively, equal to a, b, c, and d, and let $[ABCD]$ be the area endowed with a sign. We have the relation

$$[ABCD] = \tfrac{1}{2}ad\sin A + \tfrac{1}{2}bc\sin C .$$

Applying the law of cosines to triangles ABD and BCD gives

$$b^2 + c^2 - 2bc\cos C = a^2 + d^2 - 2ad\cos A .$$

It follows that

$$16[ABCD]^2 = 4a^2d^2 + 4b^2c^2 - (a^2 + d^2 - b^2 - c^2)^2 - 8abcd\cos(A + C) .$$

Replacing $\cos(A + C)$ by either

$$2\cos^2 \tfrac{1}{2}(A + C) - 1$$

O. Bottema, *Topics in Elementary Geometry*,
DOI: 10.1007/978-0-387-78131-0_13, © Springer Science+Business Media, LLC 2008

or

$$1 - 2\sin^2 \tfrac{1}{2}(A + C)$$

and introducing the notations

$$16P_1 = (-a+b+c+d)(a-b+c+d)(a+b-c+d)(a+b+c-d)$$

and

$$16P_2 = (a+b+c+d)(-a+b+c-d)(a-b+c-d)(a+b-c-d)\,,$$

we obtain, respectively,

$$[ABCD]^2 = P_1 - abcd\cos^2 \tfrac{1}{2}(A+C)$$

and

$$[ABCD]^2 = P_2 + abcd\sin^2 \tfrac{1}{2}(A+C)\,.$$

13.2

Let us consider quadrilaterals with fixed a, b, c, and d, hence also with fixed P_1. It follows from the first of these results that when it is possible to choose the quadrilateral (which needs at least five entities to be determined) in such a way that $A+C = 180°$, so that $\cos^2((A+C)/2) = 0$, then this also determines the figure for which the area is maximal. Geometrically, $A + C = 180°$ means that the quadrilateral is a *convex cyclic quadrilateral*.

The fact that there exists a convex cyclic quadrilateral among the quadrilaterals with fixed sides can be shown as follows. Let $b > d$, so that BA and

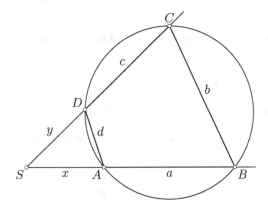

Fig. 13.2.

CD meet at S; this causes no loss of generality for the figure. If $SA = x$ and

$SD = y$, it follows from the similarity of the triangles SAD and SCB that $x : (y + c) = d : b = y : (x + a)$, whence $AD : DS : SA = (b^2 - d^2) : (ab + cd) : (ad + cb)$.

If in this proportion each number is less than the sum of the other two, the existence of triangle ADS, and therefore also that of the cyclic quadrilateral, is certain. The three inequalities can immediately be transformed into $(b - d)(a + b - c + d) > 0$ and two similar ones. These are therefore satisfied, as the constructibility of even one quadrilateral assumes that each of the four given sides a, b, c, and d is less than the sum of the remaining ones. In conjunction with Section 13.1 we therefore now have: *of all quadrilaterals with fixed sides, the convex cyclic quadrilateral has the greatest area.*

13.3

The importance of showing the existence of the convex cyclic quadrilateral for completing the proof becomes clear when we ask for the quadrilateral with the smallest area. The right-hand side of the formula

$$[ABCD]^2 = P_2 + abcd \sin^2 \tfrac{1}{2}(A + C)$$

is minimal if $\sin(A + C) = 0$, that is, if $A + C = 0$, a situation that only occurs in non-convex cyclic quadrilaterals. We may therefore conclude that if it is possible to make a non-convex cyclic quadrilateral with sides a, b, c, and d, in this order, then this is the one for which the square of the area is the smallest possible. But we immediately see that such a quadrilateral certainly does not always exist; indeed, the area would then satisfy $[ABCD]^2 = P_2$. By the definition of this last expression, this will certainly not always be a positive number.

Using reasoning similar to that of Section 13.2, we can show that there exists a non-convex cyclic quadrilateral if $P_2 \geq 0$; that is, if the sum of the longest and the shortest side is at most equal to that of the two remaining sides. Moreover, we can prove that $P_2 \leq 0$ is the necessary and sufficient condition for the existence of a quadrilateral $ABCD$ whose diagonals AC and BD are parallel, and which therefore has area *zero*. Even more basic, $P_2 \leq 0$ is the condition for the constructibility of a normal trapezoid with legs b and d and diagonals a and c. Our question now receives the following answer: if $P_2 \geq 0$, the minimum of the square of the area of the quadrilateral is equal to P_2; if $P_2 \leq 0$, the minimum is *zero*.

13.4

The most interesting result here is the property of the convex cyclic quadrilateral, partly because it is the foundation for one of the proofs given by

STEINER (1844) for the isoperimetric inequality. Let k be a curve with perimeter l. If k is not a circle (Fig. 13.3), it contains a set of four points $ABCD$ that do not lie on a circle. If we regard the area enclosed by k as being divided into

Fig. 13.3.

the quadrilateral $ABCD$ and the four remaining segments, then by replacing $ABCD$ by the larger cyclic quadrilateral with same sides and placing the segments back, we can replace the figure by one with same perimeter l and greater area. We have therefore discovered a process that to each figure with perimeter l associates one with the same perimeter and moreover, as long as the figure is not a circle, increases the area. *It follows from this that no figure other than a circle can have the maximal area.*

Once it has been determined that there exists a curve with maximal area, this must be the circle. But it is precisely this proof, which gives the most difficulty. This is no surprise if we consider the fact that the concepts of length and area and "arbitrary curve" are far from elementary. STEINER, living at a time when modern demands on the argumentation used for infinitesimal processes were not yet imposed, not only gave no proof for the existence of the maximum, but apparently did not even understand the necessity of such a proof.

13.5

The methods of STEINER have nevertheless remained significant up to now for handling isoperimetric problems (BLASCHKE, 1916 [Bla]). Let us give an example. We first note that of all triangles with fixed basis c and fixed altitude h, the isosceles triangle has the smallest perimeter. If AB is the given basis, l a line parallel to it at distance h, A' the reflection of A in l, then the variable part of the perimeter of triangle ABC is equal to $AC + CB$, hence equal to $A'C + CB$ (Fig. 13.4). It is therefore minimal if A', C, and B are collinear; in other words, if $AC = BC$. Analogously, we show that of the trapezoids $ABCD$ with fixed parallel sides AB and CD and fixed height, the isosceles trapezoid has the smallest perimeter. To do this, we draw DB' parallel to CB; then $AD + BC = AD + DB'$, and so on. Given a convex polygon such as $ABCD$ in Fig. 13.5, and an arbitrary line l in the same plane, we can carry out the following construction. Draw lines AA_1, BB', and so on

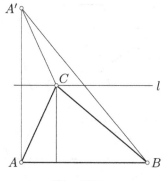

Fig. 13.4.

at the vertices, perpendicular to l, and determine the points B_1 and B_2, and so on, such that B_1B_2 is equal to BB', lies on the same line, and has its midpoint on l. This gives rise to a polygon $A_1B_1C_1D_1C_2B_2$ with the same area as $ABCD$ (it is, after all, divided up into trapezoids, respectively triangles, with the same areas as the parts of $ABCD$) and a perimeter that is at most equal to that of $ABCD$, because all trapezoids have been replaced by isosceles ones. The perimeters are equal only if $ABCD$ already possesses a symmetry axis that is parallel to l. The process outlined here is therefore

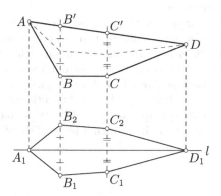

Fig. 13.5.

capable of associating, to a polygon, another one with the same area and a smaller perimeter, because there always exist lines parallel to which the polygon does not have any symmetry axis. In general, the polygon associated to an n-gon is a $(2n - 2)$-gon. We can now apply the construction to an arbitrary convex figure (Fig. 13.6). Imagine all chords PP' perpendicular to l as translated in their own direction until their midpoints lie on l. The area has not changed, on account of CAVALIERI's *principle*, while we can prove that

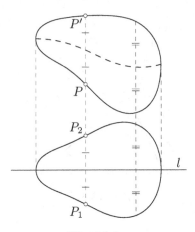

Fig. 13.6.

the perimeter has not increased. By cutting the figure up into "small" strips using lines perpendicular to l, and as such approximating the figure with a polygon, we facilitate our acceptance of these results. The line of reasoning is now as follows: assuming that for a given area there exists a figure with minimal perimeter, then it has to be the universally symmetric circle which presents this minimum. Indeed, on any other figure we can successfully apply the process mentioned above.

The fact that the circle has the greatest area of all figures with fixed perimeter is a theorem which, oddly enough, even non-mathematicians conjecture immediately. Physicists can illustrate it through a proof with a soap film. At the creation of CARTHAGE it may have influenced the actions of the astute queen DIDO who was allowed to choose a piece of land, on condition that she could enclose it with an ox hide. To obtain the greatest possible area of land, she had the hide cut up into very thin strips that were then sown together and laid out in a half-circle, with a piece of shore as diameter [Vir].

13.6

The following simple proof of the isoperimetric inequality for polygons, from which a proof for arbitrary curves follows by taking limits, was given by BOL (1943) [Bol2]. Consider the convex polygon V_1 enclosed by lines that we construct parallel to the sides of a given polygon V and at a distance a inward. In Figure 13.7, the construction has been carried out for a pentagon. For small values of a, V_1 is also a pentagon. In the figure, the situation has been drawn where V_1 becomes a quadrilateral: one of the vertices lies at the intersection point of the angle bisectors at A and B. For an even greater value of a, V_1 becomes a triangle, and so on, finally ending up at a single point, or,

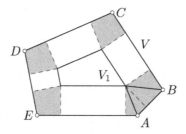

Fig. 13.7.

for example if V is a rectangle, in a line segment. Denote the areas of V and V_1 by $[V]$ and $[V_1]$, and their perimeters by p and p_1. If we slide the shaded pieces toward one another, they form a tangential polygon V_2, for which the radius of the incircle is a. Let the area be $[V_2]$, the perimeter p_2. As $[V_2] = ap_2/2$ and $[V_2] > \pi a^2$, we have $p_2 > 2\pi a$. Moreover, $[V] = [V_1] + ap_1 + [V_2]$, $p = p_1 + p_2$, and $[V_2] = ap_2/2$. It follows that

$$p^2 - 4\pi[V] = (p_1 + p_2)^2 - 4\pi([V_1] + ap_1 + \tfrac{1}{2}ap_2)$$
$$= p_1^2 - 4\pi[V_1] + (2p_1 + p_2)(p_2 - 2\pi a) > p_1^2 - 4\pi[V_1] \ .$$

The expression $p^2 - 4\pi[V]$ is therefore greater for the pentagon than for the quadrilateral. We now take a so large that a triangle appears for the first time. We can show that $p^2 - 4\pi[V]$ is greater for the quadrilateral than for the triangle. Ultimately, V_1 goes over into a point or a line segment, for which $p^2 - 4\pi[V] > 0$. The expression on the left-hand side must therefore be positive for the original polygon.

Poncelet Polygons

14.1

Let I and O be the centers of the incircle and circumcircle of triangle ABC (Fig. 14.1), let D_1 and D_2 be the midpoints of arc AB, and let E be the

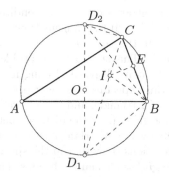

Fig. 14.1.

projection of I on BC. Then

$$\angle D_1 BI = \angle D_1 BA + \tfrac{1}{2}B = \tfrac{1}{2}C + \tfrac{1}{2}B = \angle BID_1 \,,$$

so that $D_1 I = D_1 B$ ($= D_1 A$). Moreover, triangles $D_1 D_2 B$ and ICE are similar because they have two congruent angles. We conclude that $D_1 D_2$: $D_1 B = IC : IE$; that is, $2Rr = IC \times D_1 B$, whence $2Rr = IC \times ID_1$. This last product, however, is minus the power of I with respect to circle $O(R)$, hence is equal to $R^2 - d^2$ if $OI = d$. We therefore have $d^2 = R^2 - 2Rr$ (CHAPPLE, 1746 [Cha], [Mac]; EULER, 1765 [Eul]), a relation already stated in Section 11.4, where it led to a number of inequalities in the triangle. We can also write it in the form

O. Bottema, *Topics in Elementary Geometry*,
DOI: 10.1007/978-0-387-78131-0_14, © Springer Science+Business Media, LLC 2008

$$\frac{1}{R+d} + \frac{1}{R-d} = \frac{1}{r} \, .$$

For an excircle, we find that $d_a^2 = R^2 + 2Rr_a$, and so on.

14.2

The relation is remarkable, not only because of its simplicity, but also because it contains only three quantities of the triangle, and therefore allows us to find the distance OI without determining the triangle itself. This situation is of course by no means new. After $A + B + C = 180°$, the simplest example is possibly the relation $a = 2R \sin A$ that relates side, opposite angle, and radius of circumcircle and shows these to be a non-independent threesome, unsuitable for determining a triangle.

One consequence is the following. Suppose given two circles C_1 and C_2 as well as their position with respect to each other, then without knowledge of the above theorem, we might be tempted to conjecture the existence of a solution to the problem: "Find a triangle with C_1 as circumcircle and C_2 as incircle."

Indeed, if we take A arbitrarily on C_1, the condition on C_2 first gives B and then C, fixing AB and BC. If the tangent from C to C_2 meets circle C_1 at A', we might think that the condition that A' and A coincide can be met by choosing A suitably. This should be no problem as A has the whole circumference of the circle at its disposal. Taking into account the theorem above, it turns out that this expectation is incorrect: C_1 and C_2 must be the circumcircle and incircle of a triangle, and if the condition $d^2 = R^2 - 2Rr$ is not satisfied, that is, if for given radii the distance between the centers does not have a well-determined value, no triangle is possible.

The following considerations show that in the special case that the condition is satisfied, there exist *infinitely* many solutions. There are more constructions with this property in geometry. They are known as *poristic constructions*.

14.3

Let C_1 and C_2 be the two circles (Fig. 14.2). We introduce the following notation: M_1 and M_2 are the centers of C_1 and C_2, R_1 and R_2 are their radii, $d = M_1 M_2$, N is the midpoint of $M_1 M_2$, AB and $A'B'$ are chords of C_1 that meet C_2 at D and D', and S is their intersection point. Finally, a, b, a', and b' denote the distances from A, B, A', and B' to the radical axis m. The point Q has equal powers with respect to C_1 and C_2, hence

$$(QM_1)^2 - R_1^2 = (QM_2)^2 - R_2^2 \, ,$$

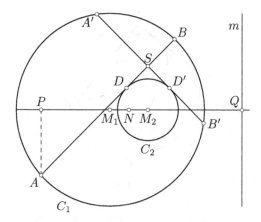

Fig. 14.2.

or

$$R_1^2 - R_2^2 = (QM_1 - QM_2)(QM_1 + QM_2) = 2d \times NQ .$$

Moreover,

$$
\begin{aligned}
AD^2 &= (AM_2)^2 - R_2^2 \\
&= (AM_2)^2 - (AM_1)^2 + R_1^2 - R_2^2 \\
&= (PM_2)^2 - (PM_1)^2 + R_1^2 - R_2^2 \\
&= 2d \times PN + 2d \times NQ \\
&= 2d \times PQ \\
&= 2da .
\end{aligned}
$$

It follows from this that

$$AS + A'S = AD + A'D' = \sqrt{2d}\left(\sqrt{a} + \sqrt{a'}\right) ,$$

and analogously

$$BS + B'S = \sqrt{2d}\left(\sqrt{b} + \sqrt{b'}\right) .$$

As the triangles $AA'S$ and $B'BS$ are similar, we also have

$$AA' : BB' = \left(\sqrt{a} + \sqrt{a'}\right) : \left(\sqrt{b} + \sqrt{b'}\right) .$$

If, conversely, we have four points A, B, A', and B' on C_1, with AB tangent to C_2, and the resulting relation holds, then $A'B'$ is also tangent to C_2, as can be shown by contradiction.

Let ABC be a triangle with C_1 as circumcircle and C_2 as incircle (Fig. 14.3), and let $A'B'C'$ be a triangle with C_1 as circumcircle such that $A'B'$ and $B'C'$ are tangent to C_2. Then

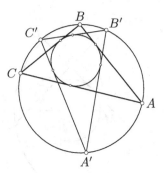

Fig. 14.3.

$$AA' : BB' = \left(\sqrt{a} + \sqrt{a'}\right) : \left(\sqrt{b} + \sqrt{b'}\right)$$

and

$$BB' : CC' = \left(\sqrt{b} + \sqrt{b'}\right) : \left(\sqrt{c} + \sqrt{c'}\right),$$

so that

$$AA' : CC' = \left(\sqrt{a} + \sqrt{a'}\right) : \left(\sqrt{c} + \sqrt{c'}\right),$$

whence it follows that $A'C'$ is tangent to C_2. We therefore have: *if there exists one triangle with C_1 as circumcircle and C_2 as incircle, then there are infinitely many such triangles. Any point of C_1 may be chosen as a vertex.*

The condition $d^2 = R^2 - 2Rr$ is not only necessary, but also sufficient for the existence of these triangles. Indeed, if we choose (Fig. 14.4) C at an intersection point of $M_1 M_2$ and C_1, A and B on C_1 and on a tangent to C_2

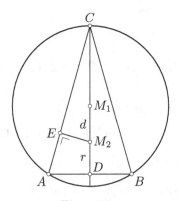

Fig. 14.4.

through D that is perpendicular to $M_1 M_2$, then

$$CD = R + r + d, \quad AD^2 = R^2 - (r + d)^2, \quad \text{and} \quad AC^2 = 2R(R + r + d).$$

For the distance x from M_2 to AC, the similar triangles CAD and CM_2E give $2Rx^2 = (R+d)^2(R-r-d)$, which reduces to $x^2 = r^2$ using the relation $d^2 = R^2 - 2Rr$. It follows that AC and BC are tangent to C_2. Thus, the one necessary triangle that we wanted has been found.

14.4

The proof given above for the existence of infinitely many triangles that are both inscribed and circumscribed as soon as one of these exists can immediately be extended to n-gons. We then obtain a theorem, which we wish to state as follows: *Given two circles C_1 and C_2, let A_1 be a point on C_1, let A_1A_2, A_2A_3, A_3A_4, ... be chords of C_1 that are tangent to C_2, and are such that A_{n+1} and A_1 coincide, so that the broken line constructed in this manner closes after n sides. Then closure also occurs if we begin at an arbitrary point of C_1. In particular, it again occurs after the n-th side.* In a slightly extended form, for two conic sections (see Section 20.6), this theorem is known as PONCELET's *porism* (1822) [Pon], [BKOR]; the elementary proof given above comes from CASEY (1858). There exists a well-known proof by JACOBI (1828) using elliptic functions, a subject of rather advanced analytic theory [Jac]. This shows, once more, how closely related certain subjects of elementary geometry are to more advanced areas of mathematics. From this proof follows an additional property of the PONCELET polygons: they do not need to be convex, closure can occur after the variable vertex has completed m revolutions along C_1. In this case, m is the same for the whole set of polygons. Based on our proof it is moreover plausible, though not proved, that two arbitrary circles do not admit PONCELET polygons, only those for which the distance d has certain values (of which there are infinitely many) depending on R_1 and R_2; d also determines n and m. Research has also been done into the existence of an *open* PONCELET polygon. If d does not have one of the special values, so that there exist infinitely many vertices A_k, these turn out to be everywhere *dense* on C_1. That is, on any arc of C_1, however small, there will, sooner or later, appear a vertex A_k.

14.5

We wish to determine the condition that the distance d must fulfill for fixed R and r to assure closure when $n = 4$. The set of quadrilaterals that are both cyclic and tangential contains an isosceles trapezoid, and it turns out to be sufficient to determine the relation for this special case. Let (Fig. 14.5) $2a$ and $2b$ $(a > b)$ denote the parallel sides. The legs are then each $a + b$. For the height $h = 2r$, we have $h^2 = (a+b)^2 - (a-b)^2$, so that $r^2 = ab$. Moreover,

$$a^2 = R^2 - (r-d)^2 \text{ and } b^2 = R^2 - (r+d)^2 \,,$$

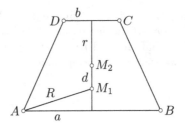

Fig. 14.5.

whence

$$r^4 = (R^2 - r^2 - d^2)^2 - 4r^2d^2 \; ,$$

or

$$\frac{1}{(R+d)^2} + \frac{1}{(R-d)^2} = \frac{1}{r^2} \; ,$$

a formula first discovered by VON FUSS (1792) [Fus].

There also exist ingenious proofs where the theorem is shown directly for an arbitrary quadrilateral with both incircle and circumcircle, sometimes called a *bicentric quadrilateral*. These quadrilaterals, which unite in them the properties of the tangential and cyclic quadrilaterals, also have other properties. We will only mention the well-known area formula $[ABCD]^2 = abcd$, which easily follows from $[ABCD]^2 = P_1$, as it is stated in Section 13.1 for a cyclic quadrilateral, and $a+c = b+d$, which is characteristic for the tangential quadrilateral.

14.6

Comparing the relations between R, r, and d for $n = 3$ and $n = 4$, which respectively state

$$\frac{1}{R+d} + \frac{1}{R-d} = \frac{1}{r} \quad \text{and} \quad \frac{1}{(R+d)^2} + \frac{1}{(R-d)^2} = \frac{1}{r^2} \; ,$$

we are inclined to make conjectures concerning the form of the relation for larger values of n. However, it is not that simple. For $n = 6$, for example, we have

$$\frac{1}{(R^2 - d^2)^2 + 4Rr^2d} + \frac{1}{(R^2 - d^2)^2 - 4Rr^2d} = \frac{1}{2r^2(R^2 + d^2) - (R^2 - d^2)^2} \; ,$$

and in general the relation becomes more complicated for larger values of n. CAYLEY (1853) gave a general process for determining the relation for arbitrary values of n [Cay].

14.7

Geometry knows many more closure problems that have the same character as the one above. That is, where there is no closure in general while in special cases closure always occurs. Consider (Fig. 14.6) a circle c and three fixed points A, B, and C. Choose a point P_0 on c, let P_1 be the intersection point

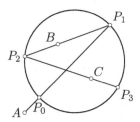

Fig. 14.6.

of P_0A and c, P_2 that of P_1B and c, and P_3 that of P_2C and c. We then can ask whether P_3 can coincide with P_0; that is, whether the broken line $P_0P_1P_2P_3$ can be closed. It turns out that for an arbitrary choice of A, B, and C, the number of points P_0 where closure occurs is equal to two, one, or zero, while it is possible, by choosing the three points in a particular manner (as the vertices of a *self-polar triangle* of the circle), to obtain closure for *every* point P_0. This problem, which can be extended to the case where not three, but $n \geq 3$ fixed points, distinct or not, are given, is known as CASTILLON's problem (1776) [Lag].

A Closure Problem for Triangles

The following closure problem is a great deal simpler than that of PONCELET
or of CASTILLON.

Let ABC be a fixed triangle, which, for the sake of convenience, we will
suppose to be acute (Fig. 15.1). We choose a point P_0 between A and B. Let

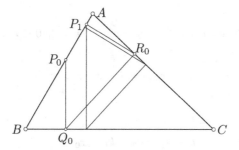

Fig. 15.1.

Q_0 be the projection of P_0 on BC, R_0 that of Q_0 on CA, and P_1 that of R_0
on AB. Taking P_1 as the origin, the construction is repeated, giving Q_1, R_1,
and P_2, and so on. P_n is the point on AB that is obtained after n revolutions.
Because of the acuteness of ABC, the points Q, R, and P lie on the sides and
not on their extensions.

We can now, for example, ask the following questions: can P_0 be chosen is
such a way that the point P_n, obtained after n revolutions, coincides with P_0,
giving a closed broken line? Another question: does there exist a limit for P_n
as n goes to ∞?

Let us mark the position of P_k using $P_k B = x_k$. With $P_0 B = x_0$, we
successively obtain

O. Bottema, *Topics in Elementary Geometry*,
DOI: 10.1007/978-0-387-78131-0_15, © Springer Science+Business Media, LLC 2008

$$BQ_0 = x_0 \cos \beta,$$
$$Q_0C = a - x_0 \cos \beta,$$
$$CR_0 = (a - x_0 \cos \beta) \cos \gamma,$$
$$R_0A = b - (a - x_0 \cos \beta) \cos \gamma,$$
$$AP_1 = \left[b - (a - x_0 \cos \beta) \cos \gamma \right] \cos \alpha,$$

and finally

$$x_1 = P_1B = c - AP_1$$
$$= c - b \cos \alpha + a \cos \alpha \cos \gamma - x_0 \cos \alpha \cos \beta \cos \gamma$$
$$= c - \cos \alpha (b - a \cos \gamma) - x_0 \cos \alpha \cos \beta \cos \gamma.$$

As $b - a \cos \gamma = c \cos \alpha$, we conclude that

$$x_1 = c \sin^2 \alpha - x_0 \cos \alpha \cos \beta \cos \gamma, \tag{15.1}$$

and in general

$$x_{n+1} = c \sin^2 \alpha - x_n \cos \alpha \cos \beta \cos \gamma.$$

With the notation

$$c \sin^2 \alpha = k, \quad \cos \alpha \cos \beta \cos \gamma = r, \quad k > 0, \ 0 < r < 1, \tag{15.2}$$

this becomes

$$x_{n+1} = k - x_n r. \tag{15.3}$$

We have

$$x_1 = k - x_0 r,$$
$$x_2 = k - x_1 r = k - kr + x_0 r^2,$$
$$x_3 = k - x_2 r = k - kr + kr^2 - x_0 r^3,$$

from which follows a formula for x_n:

$$x_n = k \left[1 - r + r^2 + \cdots + (-r)^{n-1} \right] + (-1)^n x_0 r^n$$
$$= k \frac{1 - (-r)^n}{1 + r} + (-1)^n x_0 r^n, \tag{15.4}$$

of which a rigorous proof can of course be given by induction.

Closure after one revolution is determined by the condition $x_1 = x_0$, which gives

$$x_0 = L = \frac{k}{1 + r} = \frac{c \sin^2 \alpha}{1 + \cos \alpha \cos \beta \cos \gamma}. \tag{15.5}$$

L clearly lies between A and B. All following revolutions are repetitions of the first. Closure after $n > 1$ revolutions is determined by the condition $x_n = x_0$. For $x_n = x_0$ we obtain, using (15.4),

$$\frac{1 - (-r)^n}{1 + r} [k - x_0(1 + r)] = 0 , \tag{15.6}$$

whence it follows that $x_0 = L$. The conclusion is that if closure does not occur after one revolution, it will never occur.

The limit of x_n for $n \to \infty$ follows from (15.4). As $|r| < 1$, we find $\lim x_n = L$. If we do not have $x_0 = L$ and hence have no closure, P_n tends toward this critical point L.

For an equilateral triangle with side c, we have $k = 3c/4$ and $\cos \alpha = \cos \beta = \cos \gamma = 1/2$, whence $L = 2c/3$.

16

A Class of Special Triangles

16.1

If for triangle ABC the point A_1 is the reflection of A in the opposite side BC, B_1 that of B in CA, and C_1 that of C in AB, then triangle $A_1B_1C_1$ is called the *reflection triangle* of ABC. The relationship between $A_1B_1C_1$ and ABC was studied in detail by VAN IJZEREN (1984); foremost for him was the problem of determining ABC for given $A_1B_1C_1$. This turns out to be a complicated problem; it can be reduced to solving an equation of degree seven. During his research, VAN IJZEREN came across triangles where the three reflection points A_1, B_1, and C_1 are collinear. This chapter deals with the triangles ABC with this property. We will call such a triangle with a degenerate reflection triangle an *l-triangle*. A triangle similar to an *l*-triangle is of course itself also an *l*-triangle. It follows from this that an *l*-triangle is characterized by a relation between its angles. We will use trilinear coordinates (x, y, z) (Chapter 6) with respect to ABC.

Let (Fig. 16.1) h_a be the altitude from A, then A_1 has first coordinate

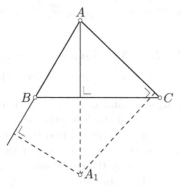

Fig. 16.1.

O. Bottema, *Topics in Elementary Geometry*,
DOI: 10.1007/978-0-387-78131-0_16, © Springer Science+Business Media, LLC 2008

$\bar{x} = -h_a$ and consequently $\bar{y} = 2h_a \cos\gamma$ and $\bar{z} = 2h_a \cos\beta$. The trilinear coordinates of A_1 are therefore $(-1, 2\cos\gamma, 2\cos\beta)$, and those of B_1 and C_1 follow from these by cyclic permutations. It follows that the reflections are collinear only when

$$\begin{vmatrix} -1 & 2\cos\gamma & 2\cos\beta \\ 2\cos\gamma & -1 & 2\cos\alpha \\ 2\cos\beta & 2\cos\alpha & -1 \end{vmatrix} = 0, \tag{16.1}$$

or after expansion,

$$-1 + 4(\cos^2\alpha + \cos^2\beta + \cos^2\gamma) + 16\cos\alpha\,\cos\beta\,\cos\gamma = 0. \tag{16.2}$$

In every triangle, we have

$$-1 + \cos^2\alpha + \cos^2\beta + \cos^2\gamma + 2\cos\alpha\,\cos\beta\,\cos\gamma = 0. \tag{16.3}$$

The following pair of equations is equivalent to the pair (16.2) and (16.3):

$$\cos^2\alpha + \cos^2\beta + \cos^2\gamma = 1\tfrac{3}{4}, \tag{16.4}$$

$$\cos\alpha\,\cos\beta\,\cos\gamma = -\tfrac{3}{8}. \tag{16.5}$$

It follows that l-triangles satisfy the (necessary) conditions (16.4) and (16.5). As there are two equations for the three angles of ABC, we can expect a one-parameter family of solutions.

16.2

From (16.5) it immediately follows that *an l-triangle is obtuse*. By convention, γ will be the obtuse angle, and $\alpha \le \beta$. Let $\cos\gamma = -p$, $0 < p < 1$. We would like to determine $\cos\alpha$ and $\cos\beta$ as functions of the parameter p. For simplicity, we write $\cos\alpha = u_1$ and $\cos\beta = u_2$, giving $u_1^2 + u_2^2 = 7/4 - p^2$ and $2u_1u_2 = 3p^{-1}/4$. We find $(u_1 + u_2)^2 = -p^2 + 7/4 + 3p^{-1}/4$ and $(u_1 - u_2)^2 = -p^2 + 7/4 - 3p^{-1}/4$, hence $u_1 + u_2 = p^{-1}\sqrt{W_1}/2$ and $u_1 - u_2 = p^{-1}\sqrt{W_2}/2$ with

$$W_1 = -4p^4 + 7p^2 + 3p = -p(p+1)(2p+1)(2p-3), \tag{16.6}$$

$$W_2 = -4p^4 + 7p^2 - 3p = -p(p-1)(2p-1)(2p+3). \tag{16.7}$$

From this we can solve $u_1 + u_2$ and $u_1 - u_2$, and therefore also u_1 and u_2, as functions of p, but we must still check whether the solutions correspond to acceptable triangles. Limiting ourselves to real triangles, we determine that W_1 is positive for every p satisfying $0 < p < 1$, but that W_2 is only positive if $2p - 1 \ge 0$, giving $\cos\gamma \le -1/2$. Hence, *in an l-triangle, the obtuse angle is greater than or equal to 120°*. In the limit case $\gamma = 120°$, we have $a = b = 30°$; in this trivial case A_1 and B_1 coincide. A second remark concerns condition (16.3). It is satisfied in every triangle. However, it only implies that $\alpha + \beta + \gamma = \pi$, where there can also be negative angles. The condition $\alpha > 0$ must also be considered. In the limit case $\alpha = 0$, the conditions give $\cos \gamma = -\cos \beta = -\sqrt{6}/4$ and $p = \sqrt{6}/4$. If we take p to be greater than $\sqrt{6}/4$ (for example, $p = \sqrt{3}/2$ with $\gamma = 150°$, $\beta = 60°$, and $\alpha = -30°$), then α is negative, because $\sqrt{6}/4$ is the only value of p with $\alpha = 0$. The conclusion is that *for every l-triangle, the obtuse angle satisfies* $120° \leq \gamma < \arccos \sqrt{6}/4 \approx 127°44'$. In such a triangle, there exists a surprisingly small margin for the obtuse angle (if φ is the upper bound, then $\cos 2\varphi = -1/4$).

The final result can be stated as follows: *the angles of an l-triangle are given by*:

$$\cos \alpha = -\tfrac{1}{4}p^{-1}(\sqrt{W_1} + \sqrt{W_2}),$$
$$\cos \beta = -\tfrac{1}{4}p^{-1}(\sqrt{W_1} - \sqrt{W_2}), \qquad (16.8)$$
$$\cos \gamma = -p,$$

where W_1 and W_2 are determined by (16.6) and (16.7) and $1/2 \leq p < \sqrt{6}/4$.

A numeric example (J.M. BOTTEMA, 1985) is given in Figure 16.2, where

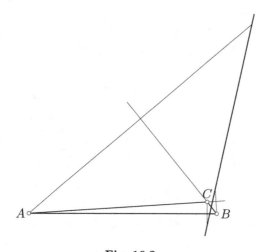

Fig. 16.2.

the value for γ has been chosen to be $126°$. It follows that $p = 0.58779$, and using (16.6), (16.7), and (16.8) that $\alpha = 3.7°$ and $\beta = 50.3°$.

16.3

Another property of l-triangles came to light by chance. It started with the question (GUIMARD, 1985) whether in a triangle, the center N of the nine-point circle can lie on the circumcircle. For this we must look on the EULER line (Chapter 4) on which, in addition to N, the points H, G, and O also lie. Moreover, we have $HN : NG : GO = 3 : 1 : 2$; that is, $NO = (1/2)HO$. Trigonometry provides the formula

$$HO^2 = (1 - 8\cos\alpha\,\cos\beta\,\cos\gamma)R^2\ .$$

In our case, where $NO = R$, this gives $\cos\alpha\,\cos\beta\,\cos\gamma = -3/8$, which is exactly our relation (16.5).

The conclusion is that *in an l-triangle, the center of the nine-point circle lies on the circumcircle, and vice versa.*

The true geometer will not be satisfied by the given proof, which uses the correspondence of the characteristic trigonometric relations for the two triangles. He can obtain help from G.R. VELDKAMP (1986) and his fine theorem (whose proof will not be given here): *the reflection triangle of ABC is similar to the pedal triangle of ABC for N* [Vel]. The argument is then completed by calling on WALLACE's theorem (Chapter 9).

17

Two Unusual Conditions for a Triangle

17.1

In 1967, people became interested in triangles with the property that *the center of mass of the perimeter* lies on the *incircle*. This somewhat odd problem is justified by the fact that for this type of triangle there exists a simple and even elegant relation between the sides. This was reason enough (J.T. GROENMAN, 1989) to try to deduce the said relation through direct algebraic computations using barycentric coordinates, a procedure that provides a proof not only of the computing talents of the person involved but also of his perseverance.

Let ABC be the triangle, $M(\bar{x}, \bar{y}, \bar{z})$ the center of mass of the perimeter, and h_a the altitude from A. Then

$$\bar{x} = \frac{b\frac{1}{2}h_a + c\frac{1}{2}h_a}{a+b+c} = \frac{(b+c)[ABC]}{a(a+b+c)} \, ,$$

so that in homogeneous barycentric coordinates,

$$M = \big((b+c), (c+a), (a+b)\big) . \tag{17.1}$$

If D, E, and F are the points where the incircle touches, respectively, BC, CA, and AB, then for D, we have: $\bar{x} = 0$, $\bar{y} = (s-c)\sin\gamma$, and $\bar{z} = (s-b)\sin\beta$, whence, in homogeneous barycentric coordinates,

$$D = \big(0, (a+b-c), (a-b+c)\big) . \tag{17.2}$$

Analogous answers hold for E and F.

The incircle is tangent to $X = 0$ at D, and so on, and after elementary computations, we find the following equation for the incircle:

$$(-a+b+c)^2 X^2 + (a-b+c)^2 Y^2 + (a+b-c)^2 Z^2$$
$$-2(-a+b+c)(a-b+c)XY - 2(a-b+c)(a+b-c)YZ \tag{17.3}$$
$$-2(a+b-c)(-a+b+c)ZX = 0 \, .$$

O. Bottema, *Topics in Elementary Geometry*,
DOI: 10.1007/978-0-387-78131-0_17, © Springer Science+Business Media, LLC 2008

The desired relation is obtained by substituting (17.1) in (17.3). This is a laborious task, which, via the transitional state

$$2(b^2c^2+c^2a^2+a^2b^2)+a^3(b+c)+b^3(c+a)+c^3(a+b) = 5abc(a+b+c)\,, \quad (17.4)$$

leads to the well-earned final result

$$\frac{1}{a} + \frac{1}{b} + \frac{1}{c} = \frac{10}{a+b+c}\,. \tag{17.5}$$

Whoever is worried that the set of triangles satisfying (17.5) might be empty can be reassured by the example of the very real triangle $a = 1$, $b = c = 2$.

17.2

A related question states: *in which triangles does the Nagel point lie on the incircle?* (GROENMAN, 1989 [Gro]). As can be deduced from Section 2.6, in homogeneous barycentric coordinates $Na = (s - a, s - b, s - c)$. After substituting this in (17.3), some computation gives

$$(s - a)^4 + (s - b)^4 + (s - c)^4$$
$$- 2(s - b)^2(s - c)^2 - 2(s - c)^2(s - a)^2 - 2(s - a)^2(s - b)^2 = 0\,. \quad (17.6)$$

The left-hand side of (17.6) is an expression that is well known from the formula for the area of a triangle. It may be decomposed into four linear factors:

$$\big((s - a) + (s - b) + (s - c)\big)\big(-(s - a) + (s - b) + (s - c)\big)$$
$$\times \big((s - a) - (s - b) + (s - c)\big)\big((s - a) + (s - b) - (s - c)\big) = 0\,,$$

that is,

$$s(s + a - b - c)(s - a + b - c)(s - a - b + c) = 0,$$

or finally

$$(3a - b - c)(3b - c - a)(3c - a - b) = 0. \tag{17.7}$$

The desired triangles are therefore those where the sum of two of the sides is three times the third side.

A numerical example is the well-known triangle $a = 3$, $b = 4$, $c = 5$.

A Counterpart for the Euler Line

On the EULER line e (Chapter 4) lie the classical special points: the centroid G, the orthocenter H, and the center O of the circumcircle. The center of the nine-point circle also lies on e. Using barycentric coordinates, we will now deduce another set of three collinear points associated to a triangle ABC.

The centroid satisfies $G = (1, 1, 1)$. The trilinear coordinates of the center I of the incircle are equal to each other, whence $I = (a, b, c)$. Our third point is the NAGEL point Na defined in Section 2.6. If (Fig. 2.5, page 11) A' on BC is the point where the corresponding excircle touches BC, then $BA' = s - c$ and $CA' = s - b$. The equation of the cevian AA' is therefore $Y : Z = (s-b) : (s - c)$. The NAGEL point Na therefore satisfies $Na = (s - a, s - b, s - c)$. The coordinates of the three points G, I, and Na are the elements of the determinant

$$d = \begin{vmatrix} 1 & 1 & 1 \\ a & b & c \\ s - a & s - b & s - c \end{vmatrix} . \tag{18.1}$$

After expansion, or, even quicker, by adding the second line to the third one, it turns out that $d = 0$. The conclusion is that *in every triangle the centroid, the center of the incircle, and the* NAGEL *point are collinear*. The equation for the line f joining them follows from (18.1):

$$(b - c)X + (c - a)Y + (a - b)Z = 0 . \tag{18.2}$$

The barycentric coordinates of the center of mass M of the perimeter have been deduced in Section 17.1:

$$M = (b + c, c + a, a + b) ,$$

and we immediately see that M satisfies (18.2). The conclusion is that *M also lies on the line f passing through G, I, and Na.*

The deduction of the line f once more shows the power of the analytic method, in particular if we compare it to the usual complicated geometric proof (JOHNSON, 1929 [Joh2]).

O. Bottema, *Topics in Elementary Geometry*,
DOI: 10.1007/978-0-387-78131-0_18, © Springer Science+Business Media, LLC 2008

The position of the three collinear points G, I, and Na with respect to one another can also be determined. For this, we first normalize their homogeneous coordinates. Taking the area of the triangle ABC as unity gives the coordinates $G = (1/3, 1/3, 1/3)$, $I = (a/2s, b/2s, c/2s)$, and $Na = ((s-a)/s, (s-b)/s, (s-c)/s)$. It follows from this that the distances from these points to side BC are in the proportions $1/3 : a/2s : (s-a)/s$. If in each point, we draw the line parallel to BC, we find

$$NaG : GI = \left(\frac{s-a}{s} - \frac{1}{3}\right) : \left(\frac{1}{3} - \frac{a}{2s}\right) = \frac{2s - 3a}{3s} : \frac{2s - 3a}{6s} = 2 : 1 ,$$

strangely enough the same proportion that also holds for the EULER line e, where we let G correspond to G, Na to H, and I to O.

Menelaus's Theorem; Cross-Ratios and Reciprocation

19.1

Let l be a line (Fig. 19.1) that meets the sides BC, CA, and AB of triangle ABC at, respectively, P, Q, and R, and let us draw an auxiliary line

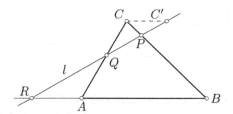

Fig. 19.1.

through C that is parallel to AB and meets l at C'. Then it follows from the similarity of the triangles CPC' and BPR that $CC' : RB = PC : BP$, and from that of the triangles RAQ and $C'CQ$ that $AR : CC' = AQ : CQ$. After multiplication this gives

$$\frac{AR}{RB} \times \frac{BP}{PC} \times \frac{CQ}{QA} = -1 .$$

This equality, known as MENELAUS's theorem, gives the relation between the positions of three points on the sides of a triangle, when these points are collinear [Men]. In a similar manner as in CEVA's theorem, which is its counterpart, we can show that the relation is not only necessary but also sufficient for the collinearity. The theorem also holds, after the extension of certain notions, for a line parallel to a side. We can use it to show that three points are collinear. In Chapter 20, we will come across a number of examples of this application.

O. Bottema, *Topics in Elementary Geometry*,
DOI: 10.1007/978-0-387-78131-0_19, © Springer Science+Business Media, LLC 2008

19.2

If four concurrent lines a, b, c, and d are met by lines l_1 and l_2 at, respectively, A_1, B_1, C_1, and D_1 and A_2, B_2, C_2, and D_2 (Fig. 19.2), and we draw the

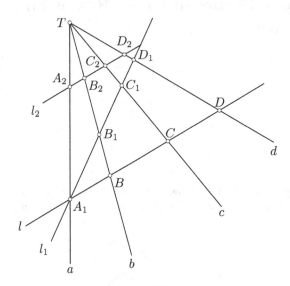

Fig. 19.2.

line parallel to l_2 through A_1, giving additional intersection points B, C, and D, then the application of MENELAUS's theorem to triangle A_1B_1B with c respectively d as intersecting line gives

$$\frac{A_1C_1}{C_1B_1} \times \frac{B_1T}{TB} \times \frac{BC}{CA_1} = -1$$

and

$$\frac{A_1D_1}{D_1B_1} \times \frac{B_1T}{TB} \times \frac{BD}{DA_1} = -1,$$

so that

$$\frac{A_1C_1}{B_1C_1} \times \frac{B_1D_1}{A_1D_1} = \frac{A_1C}{BC} \times \frac{BD}{A_1D}.$$

The last expression is clearly equal to

$$\frac{A_2C_2}{B_2C_2} \times \frac{B_2D_2}{A_2D_2}.$$

If A, B, C, and D are four points on a line, the expression

$$\frac{AC \times BD}{BC \times AD},$$

which can also be written as

$$\frac{AC}{BC} : \frac{AD}{BD},$$

that is, as the ratio of two ratios, is called the *cross-ratio of ABCD*.

We have thus determined the remarkable fact that in the case of a *perspective projection* (that is, the construction giving rise to $A_2B_2C_2D_2$ by joining A_1, B_1, C_1, and D_1 to the *center* T and intersecting with the image line l_2) the cross-ratio remains unchanged, or *invariant*. In the case of the less radical parallel projections this is already true for the usual ratios of two line segments.

The cross-ratio $d = (ABCD)$ depends on the order of the points, but even then takes on only 6 different values for the 24 possible *permutations*, as $(ABCD) = (BADC)$, and so on. If the points are all distinct, then $d \neq 0, 1$, and d is also unequal to ∞. Conversely, if d is a number other than 0 or 1, then for fixed distinct A, B, and C, there always exists a unique point D for which $(ABCD) = d$. The case $d = -1$ is special; for the 24 permutations, the cross-ratio only takes on 3 values, namely, -1, 2, and $1/2$. It is also possible to choose a value for d where this number is only 2, but as that value is not real, that case is of no importance for our geometry.

If $(ABCD) = -1$, we say that C and D are *harmonic conjugates* with respect to A and B. We may exchange A and B, and also C and D, and even the pair AB with the pair CD. For example, if the bisectors of angle C of triangle ABC meet side AB at S_1 and S_2, then S_1 and S_2 are harmonic conjugates with respect to A and B.

If point A of the quadruple $ABCD$ lies at infinity, then $(ABCD)$ is none other than the ratio $BD : BC$. The theorem above still holds for such a case; l_2 is then for example parallel to a. If A and B are harmonic conjugates with respect to C and D and A is the point at infinity, then B is the midpoint of CD.

19.3

If A, B, C, and D (Fig. 19.3) are four points in general position, then together with the six line segments joining them pairwise they form a figure called a *complete quadrangle*. If P, S_1, and S_2 are the intersection points of the pairs of *opposite* sides AB and CD, AC and DB, and AD and BC, and Q and R are the intersection points of S_1S_2 and AB and of S_1S_2 and CD, then the projection of $CDPR$ from S_1 onto AB gives the figure $BAPQ$. It follows from this that $(ABPQ) = (BAPQ)$. These cross-ratios are therefore equal to their own inverses; that is, P *and* Q *are harmonic conjugates with respect to* A *and* B, *and* P *and* R *are harmonic conjugates with respect to* C *and* D.

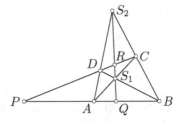

Fig. 19.3.

19.4

Let us choose a point P on the diameter AB of circle M (Figs. 19.4 and 19.5) and consider a line through P that meets the circle at C and D. Let H and S be the intersection points of AC and BD and of AD and BC, P' and Q those of SH and AB and of SH and CD. As AC lies perpendicular to BC and BD lies perpendicular to AD, H is the orthocenter of triangle ABS. The segment SH is therefore perpendicular to AB, intersecting it at the point P', which together with P forms a pair of harmonic conjugates with respect to A and B, and is therefore independent of the manner in which PCD may be varied.

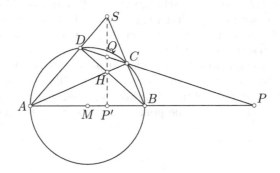

Fig. 19.4.

SH is called the *polar p* of P with respect to the circle. It contains all points Q on the lines PCD that together with P form a pair of harmonic conjugates with respect to C and D. If P lies outside the circle, as in Fig. 19.4, then p intersects the circle, meeting it at the points where the tangents from P touch it. It is now easy to see that each line p is the polar of a single point, the *pole* of the line. If P_1 lies on the polar of P_2, then P_2 lies on the polar of P_1. The polar of a point on the circle is the tangent line at that point.

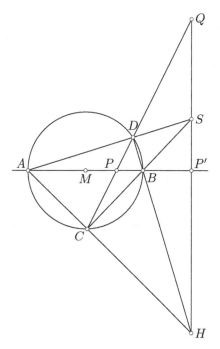

Fig. 19.5.

19.5

Every point has a polar except for the center M of the circle. M is the center of every chord through M, its harmonic conjugate therefore lies infinitely far away, so that it seems natural to regard the "line at infinity" as the polar of M. Every line has a pole except for a line m through M. The properties of the correspondence are preserved if we let the point at infinity in the direction perpendicular to m correspond to m.

Having taken care of infinity, we have a *duality principle* for this reciprocation with respect to a circle. To every point P, without exception, it associates a line p and conversely to every line p it associates a point P. Moreover, the following properties hold: if a number of points P_1, P_2,... lie on a line l, then their polars p_1, p_2,... concur at a point L, the pole of l; and if a number of lines concur, then their poles are collinear. If we have a theorem concerning only points and lines and the position of certain points on certain lines (and therefore not distances, angles, and so on), then by subjecting the figure to our reciprocation, we can deduce a new theorem whose statement can be obtained from the given one by exchanging point with line, point on a line with line through a point, and line segment joining two points with intersection point of two lines. The new theorem is called the *dual theorem* of the given one. A number of examples will follow in Chapter 20.

The Theorems of Desargues, Pappus, and Pascal

20.1

Consider two triangles $A_1B_1C_1$ and $A_2B_2C_2$ such that A_1A_2, B_1B_2, and C_1C_2 meet at O (Fig. 20.1). If P, Q, and R are the intersection points of B_1C_1

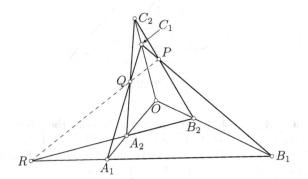

Fig. 20.1.

and B_2C_2, of C_1A_1 and C_2A_2, and of A_1B_1 and A_2B_2, then we can apply MENELAUS's theorem to triangle B_1C_1O with intersecting line B_2C_2, whence it follows that

$$\frac{OB_2}{B_2B_1} \times \frac{B_1P}{PC_1} \times \frac{C_1C_2}{C_2O} = -1 .$$

Analogous relations arise by considering triangle C_1A_1O with line C_2A_2 and triangle A_1B_1O with A_2B_2. Multiplying the corresponding parts of the three equations gives

$$\frac{B_1P}{PC_1} \times \frac{C_1Q}{QA_1} \times \frac{A_1R}{RB_1} = -1 ,$$

whence it follows that P, Q, and R are collinear. We have therefore proved DESARGUES's theorem (1639): *if the lines joining the corresponding vertices*

O. Bottema, *Topics in Elementary Geometry*,
DOI: 10.1007/978-0-387-78131-0_20, © Springer Science+Business Media, LLC 2008

*of two triangles are concurrent, the intersection points of the pairs of corre-
sponding sides are collinear* [Bos].

Special cases that arise when parallel lines occur in the figure can in general
be dealt with by introducing points *at infinity* all lying on the *line at infinity*
mentioned before.

20.2

Let six points $A_1A_2A_3A_4A_5A_6$ lie in arbitrary order on a circle (Fig. 20.2).
Let P be the intersection point of A_1A_2 and A_4A_5, Q that of A_2A_3 and A_5A_6,

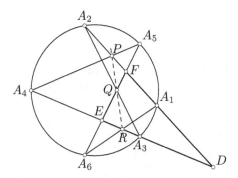

Fig. 20.2.

and R that of A_3A_4 and A_6A_1. Furthermore, let D be the intersection point
of A_1A_2 and A_3A_4, E that of A_3A_4 and A_5A_6, and F that of A_5A_6 and A_1A_2.

We apply MENELAUS's theorem three times to triangle DEF, namely, with
intersecting lines A_2A_3, A_4A_5, and A_6A_1. This gives

$$\frac{DA_3}{A_3E} \times \frac{EQ}{QF} \times \frac{FA_2}{A_2D} = -1$$

and two analogous relations. Multiplying the corresponding parts and using
$DA_1 \times DA_2 = DA_3 \times DA_4$ and analogous relations for E and F gives

$$\frac{DR}{RE} \times \frac{EQ}{QF} \times \frac{FP}{PD} = -1 ,$$

which shows that P, Q, and R are collinear. This shows the following theorem,
proved by PASCAL in 1640 at the age of sixteen: *the intersection points of the
three pairs of opposite sides of a hexagon inscribed in a circle are collinear*
[Pas].

The theorem also holds if two successive vertices coincide and the corresponding line joining them is taken to be the tangent line to the circle. *The*

same theorem holds for a hexagon inscribed in a pair of lines (PAPPUS, around 300 AD [Pap]). If A_1, A_3, and A_5 lie on a line l_1, and A_2, A_4, and A_6 lie on a line l_2, we apply MENELAUS's theorem to triangle DEF, once more with intersecting lines A_2A_3, A_4A_5, and A_2A_1, but now also with intersecting lines l_1 and l_2 (Fig. 20.3). After a simple elimination we again find that P, Q, and R are collinear.

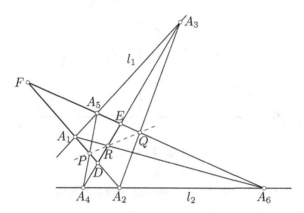

Fig. 20.3.

20.3

A *complete quadrilateral* is the figure made up of four lines l_1, l_2, l_3, and l_4 in general position. It has six vertices: the intersection points of the sides taken pairwise. To each vertex we associate the opposite vertex, namely, the one that arises by intersecting the two sides on which it does not lie. The line joining two opposite vertices is not a side; it is called a *diagonal*. The following theorem, due to GAUSS (1810), holds for the resulting figure: *the midpoints of the diagonals of a complete quadrilateral are collinear* [Gau]. Let A_{ij} denote the intersection point of l_i and l_j, and M_1, M_2, and M_3 the midpoints of $A_{12}A_{34}$, $A_{13}A_{42}$, and $A_{14}A_{23}$ (Fig. 20.4). Applying MENELAUS's theorem to triangle $A_{12}A_{23}A_{31}$ with intersecting line l_4 gives

$$\frac{A_{12}A_{24}}{A_{24}A_{23}} \times \frac{A_{23}A_{34}}{A_{34}A_{31}} \times \frac{A_{31}A_{14}}{A_{14}A_{12}} = -1 \, .$$

If P_1, P_2, and P_3 are the midpoints of $A_{12}A_{31}$, $A_{23}A_{12}$, and $A_{31}A_{23}$, then M_1 lies on P_1P_2, M_2 on P_3P_1, and M_3 on P_2P_3, while $P_1M_2 = A_{12}A_{24}/2$, and so on. It follows that

$$\frac{P_1M_2}{M_2P_3} \times \frac{P_3M_3}{M_3P_2} \times \frac{P_2M_1}{M_1P_1} = -1 \, ,$$

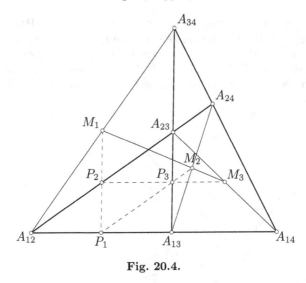

Fig. 20.4.

whence the theorem follows by considering triangle $P_1 P_2 P_3$.

20.4

Of the four theorems mentioned above, those of DESARGUES and of PAPPUS are the simplest as both in the result and in the data only the most fundamental notions of plane geometry appear, namely, those of a *point*, a *line*, and a *point lying on a line*. In PASCAL's theorem, the already much more complicated concept of a circle appears, while in GAUSS's the *midpoint* of two points is mentioned.

Any property of a plane figure F, also holds for every figure congruent to F, and unless it concerns numerical measures, even for every figure similar to F. The following may sound unnecessarily weighty to those who are not knowledgeable on this subject and who may be inclined to consider the remark as a triviality, but it has a deep meaning. Namely, our geometric statements are invariant with respect to those modifications which we may bring to the figures by giving them another position and possibly drawing them on another scale; that is, they are invariant for rigid motions and for similarities.

We can also transform a figure in another way. For example, we can project the whole plane of our plane geometry by parallel projection onto another plane V' in space, and put V' back in V in some position or another. Can our notions and theorems also withstand such interventions? Apparently the answer is that some can, while some cannot.

Those that *cannot* include: the notion of a right angle, because the projection of a right angle is in general not a right angle; hence also the Pythagorean theorem and all theorems deduced from it; also anything concerning a circle, such as PASCAL's theorem in the form given above.

But the following *can*: the notion of a point, and of a line, and of a point lying on a line, and also: the parallelism of lines, the ratio between two line segments on the same or on parallel lines, the midpoint of two points, GAUSS's theorem.

Together, the notions and theorems that are invariant under the transformations mentioned above, called affine, form the so-called *affine geometry*. More important is the geometry that arises by not only taking parallel projections from V onto V', but also *central* projections. The notions of a point and a line are still invariant, at least if we no longer make a distinction between "normal" points and those "at infinity", as are the notion of "a point lying on a line" and, as we have proved, the ratio of two ratios of line segments on the same line. The same holds for DESARGUES's theorem and PAPPUS's theorem. These therefore have a very primitive character, as they belong to what is called *projective geometry*, a chapter of mathematics developed in the nineteenth century. Let us note that we can also define affine and projective transformations without using an auxiliary plane V'.

20.5

We have shown the theorems of DESARGUES and PAPPUS using MENELAUS's theorem, which belongs to affine geometry, and clearly not to projective geometry. So how does a projective geometer, who wishes to stay within his system, prove these theorems? For DESARGUES's theorem this is possible using spatial considerations. If in Figure 20.1, we regard OA_1, OB_1, and OC_1 as being three non-coplanar lines, hence $A_1B_1C_1$ and $A_2B_2C_2$ as being two planes, then the theorem in fact says that the locus of the points common to both planes is a line.

The question whether a proof that restricts itself to the plane and only uses the axioms of the joining of points and the intersection of lines is possible was answered negatively by the research of HILBERT (1899) [Hil]; the same holds for PAPPUS's theorem.

We therefore include both theorems in the set of axioms of projective geometry, all the more because this geometry can then essentially be developed without any other axioms. With this, the great theoretical significance of both theorems is sufficiently illustrated. Let us also remark that in 1905, HESSENBERG showed, through an ingenious proof, that DESARGUES's theorem is a consequence of that of PAPPUS, so that it suffices to introduce the latter theorem as an axiom [Hes]. Conversely, it has been shown that PAPPUS does *not* follow from DESARGUES.

20.6

Parallel projection transforms a circle into a curve that is called an *ellipse*. Central projection transforms it into the intersection of a (double) cone with a plane that does not go through its apex. Such an intersection is called a *conic section* and can be an ellipse, a *parabola*, or a *hyperbola*. It is now clear that PASCAL's theorem holds for every conic section, and that the theorem, stated in this generality, is part of projective geometry.

20.7

As the elements at infinity are assimilated to the others in projective geometry, which is not the case in the usual and in affine geometry, it follows from Section 19.5 that the *duality principle* holds. Consequently, each theorem leads to a new one by interchanging point and line. We easily see that DESARGUES's theorem is self-dual. The dual of PAPPUS's hexagon theorem states (Fig. 20.5): if the lines a_1, a_3, and a_5 meet at L_1, and a_2, a_4, and a_6 meet

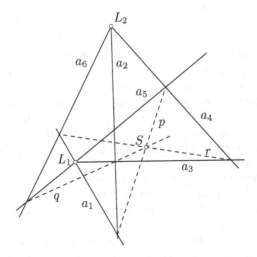

Fig. 20.5.

at L_2, and S_{ij} is the intersection point of a_i and a_j, then the lines joining S_{12} and S_{45}, S_{23} and S_{56}, and S_{34} and S_{61} are concurrent. PASCAL's theorem has an analogous dual, where a_1, a_2, a_3, a_4, a_5, and a_6 are tangents to a conic section (BRIANCHON, 1806 [Bri]).

20.8

In Figure 20.3, the illustration of PAPPUS's theorem, there are nine points A_i, P, Q, and R and nine lines such that three lines go through each point, and each line goes through three points. Such a figure is called a *configuration*. The definition does not include that the points and lines must be equal in number. The name is also used to describe a figure consisting of a points and b lines, where n lines go through each point and m points lie on each line (where, as we can easily see, $an = bm$ must hold). Such a figure is denoted by (a_n, b_m). A *complete n-gon* is a configuration $(n_{n-1}, (n(n-1)/2)_2)$.

In a symmetric configuration $n_3 = (n_3, n_3)$, n must be at least 7, because three concurrent lines also contain 3×2 other points. The configuration 7_3 exists from a combinatorial point of view, but it cannot be realized geometrically. The same holds for 8_3, at least if we restrict ourselves to figures without any points at infinity.

On the other hand, there are *three* different configurations 9_3, namely, PAPPUS's configuration and those of Figures 20.5a and 20.5b, which are less important. The difference between the three configurations has nothing to do

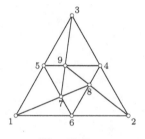

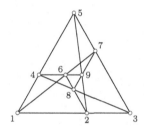

Fig. 20.5a.   **Fig. 20.5b.**

with form or dimension; it is a difference in *structure*. Figure 20.1 is also a configuration, namely, a 10_3. The golden age of research in the field of configurations was around 1900; Dutch geometers (JAN DE VRIES [Vri1], [Vri2], BARRAU [Bar]) made important contributions.

Inversion

21.1

A transformation that is more important for elementary geometry than any affine or projective one, is *inversion*. If O is the *center* and the number m is the *power* of the inversion $O(m)$, then to an arbitrary point P, this associates a point P' that lies on OP in such a way that $OP \times OP' = m$. No point is associated to O itself. We can take care of any objections originating from this situation by introducing a point L said to be *at infinity*. This closure of the plane, which is completely different from the one that takes place in projective geometry, turns out to generate a completely satisfactory geometric system. Every line in the plane is assumed to go through L. Two lines therefore, in general, have two common points, which coincide at L if the lines are parallel.

21.2

To a line through O, an inversion $O(m)$ apparently associates the line itself. If l

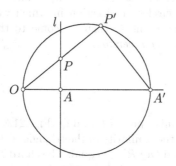

Fig. 21.1.

O. Bottema, *Topics in Elementary Geometry*,
DOI: 10.1007/978-0-387-78131-0_21, © Springer Science+Business Media, LLC 2008

does not pass through O, A' is the inverse point of the projection A of O on l (Fig. 21.1), and P' that of an arbitrary point P on l, then $m = OA \times OA' = OP \times OP'$, so that $OA : OP = OP' : OA'$. The triangles OAP and $OP'A'$ are therefore similar, and the angle P' is a right angle. It follows from this that the locus of the P' is a circle through O. Conversely, the inverse figure of a circle through O is a line that does not pass through O. Two parallel lines invert into two circles tangent to each other at O.

If circle c does not pass through O (Fig. 21.2), P is an arbitrary point,

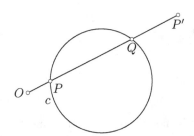

Fig. 21.2.

P' its inverse point, and Q the second intersection point of OP with c, then $OP \cdot OP' = m$. Moreover, $OP \cdot OQ$ is also constant, namely, equal to the power k of O with respect to c. It follows that $OP' = mOQ/k$. For the whole circle c, inversion therefore is the same as a central dilation with center O and scaling factor m/k. Consequently, the inverse figure is a circle. It can even be c itself if $k = m$. All circles with respect to which the power of O is equal to the power of the inversion are therefore invariant. If m is positive, there also exist invariant points, namely, those on the circle with center O and radius $\sqrt{m}$. We immediately see that the invariant circles are those that meet this circle perpendicularly.

It seems as if inversion has a very disorganizing influence on our figures, as it does not even preserve straightness. If, however, we decide to regard straight lines as being circles, a direction in which we already took a step when we introduced the point L, then we come to the positive statement: *circles remain circles under inversion.*

21.3

Let k be a curve, containing points P and Q (Fig. 21.3), and let k', P', Q' be the inverse figure, then the similarity of the triangles OPQ and $OQ'P'$ implies that the angles OPQ and $OQ'P'$ are equal. The chord PQ therefore makes an angle with OP that is equal to the one that the chord $P'Q'$ makes with OQ, even if $P'Q'$ is not, as a whole, the inverse of PQ. If Q nears P, PQ nears

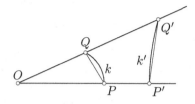

Fig. 21.3.

the tangent line to k at P, Q' nears P', $P'Q'$ nears the tangent line to k' at P', and OQ nears OP, which shows that the tangent lines to k and k' at the points P and P' make equal angles with OP. The same holds for a second curve l going through P and the associated curve l' through P'. By subtracting corresponding angles from each other, we obtain: *the angle that two curves make at an intersection point P is equal to the one made by the inverse curves at the inverse point P'.*

Inversion therefore preserves the size of angles; it is an *angle-preserving* or *conformal* transformation. In particular, it preserves *tangency* and *perpendicular intersection.*

21.4

Inversion originated in the middle of the nineteenth century and was first researched extensively by LIOUVILLE (1847) [Lio]. Its great importance for elementary geometry is clear if we consider that it makes it possible to transform certain exercises in which circles are concerned , and in particular many constructions, into less complicated ones where one or more circles have been replaced by a line. For similar reasons, inversion was soon applied by physicists, for example by THOMSON in the theory of electric fields [Tho1], [Tho2]. The transformation is also important from a more theoretical point of view. In analogy with what we have seen for affine and projective geometry, a *conformal geometry* or *inversive geometry* was developed, which only studies such notions and properties that are not only invariant for rigid motions and similarities, but also for inversions. This geometry therefore includes the notions of circle and angle, but not that of line, radius, or center.

The figure of a triangle, that is, of three points, is not interesting in this geometry. We can in fact prove that it is always possible to choose an inversion in such a way that three given points are mapped into three other given points, so that from the point of view of conformal geometry all triangles are "congruent". This is clearly not the case for quadrilaterals, since four points can either all lie on a circle, or not. It is then no coincidence that we will use inversion to prove certain properties of quadrilaterals: these are in fact theorems from conformal geometry.

22

The Theorems of Ptolemy and Casey

22.1

Inversion does not preserve the distance between two points. This also holds
for parallel projection and central projection. However, in those cases certain
expressions, namely, the ratio of two collinear distances, respectively the ratio
of two such ratios, are preserved. Something similar holds for inversion. Let
A' and B' be the inverse points of A and B for the inversion $O(m)$, then the
similarity of the triangles OAB and $OB'A'$ implies that

$$AB : A'B' = OA : OB \,,$$

a proportion that is equal to $(1/m) \times OA \times OB$. Let $ABCD$ be four points,
$A'B'C'D'$ the inverse points, then

$$AB \times CD : A'B' \times C'D' = \frac{1}{m^2} \times OA \times OB \times OC \times OD \,.$$

It follows that an inversion scales the three products $p_1 = AB \times CD$, $p_2 = AC \times DB$, and $p_3 = AD \times BC$ by the same factor. *The ratios between these
three quantities are therefore invariant under inversion.* They characterize
quadrilateral $ABCD$.

22.2

These ratios allow more than one simple geometric interpretation. Consider
an inversion $D(m)$ and let A', B', and C' be the inverse points of A, B, and C,
then $AB : A'B' = (1/m) \times DA \times DB$, hence $A'B' = m \times (AB \times DC)/(DA \times DB \times DC)$, and so on. Therefore, *if we invert the quadrilateral with respect to
one of its vertices, the inverses of the vertices that lie in the finite plane form
a triangle, of which the sides are in the same proportions as p_1, p_2, and p_3.*

O. Bottema, *Topics in Elementary Geometry*,
DOI: 10.1007/978-0-387-78131-0_22, © Springer Science+Business Media, LLC 2008

It immediately follows that *none of the three products is greater than the sum of the other two*.

One of the products is *equal* to the sum of the other two if and only if the points A', B', and C' arising through the inversion $D(m)$ are collinear. This occurs if the circumcircle of triangle ABC passes through the center of the inversion, that is, through D; in other words, if $ABCD$ is a *cyclic quadrilateral*. If the vertices lie in the order $ABCD$, then B' lies between A' and C', so that $A'C' = A'B' + B'C'$, that is, $AC \times BD = AB \times CD + BC \times AD$. Conversely, if this relation holds, then $A'B'C'$ is a straight line and $ABCD$ is a cyclic quadrilateral. This is PTOLEMY's theorem (ca. 150 AD) [Pto], [Too].

The proof without inversion, which is well known, is based on the ingenious use of two auxiliary lines that meet on a diagonal and form two pairs of similar triangles. That the converse also holds is most easily proved by also drawing these auxiliary lines in an arbitrary quadrilateral, noting that, in general, these do not meet on a diagonal, and deducing from this an inequality between the three products. More precisely,

$$AC^2 \times BD^2 = AB^2 \times CD^2 + BC^2 \times AD^2 - 2AB \times CD \times BC \times AD \cos(A+C) ,$$

an extension of the law of cosines. We find PTOLEMY's relation both for convex $(A+C = 180°)$ and non-convex cyclic quadrilaterals $(A+C = 0°)$. The relation between this and the trigonometric addition formulas is also known. Through a judicious choice of signs for each of the three products, PTOLEMY's theorem can also be written in the symmetric form $p_1 + p_2 + p_3 = 0$.

22.3

While at least four points were necessary to obtain a figure with an invariance under inversion, there exists another figure with a conformal invariant, namely, *the figure of two circles*. For two intersecting circles this is clear: they intersect under a certain angle φ that is invariant under inversion. If r_1 and r_2 are the radii of the circles, d the distance between their centers, the law of cosines gives

$$\cos \varphi = k = \frac{r_1^2 + r_2^2 - d^2}{2r_1 r_2} .$$

The expression for k can be written for two arbitrary circles, even if they do not meet; in that case, k is not the cosine of a real angle. It is natural to conjecture that in that case k is also preserved by inversion. Indeed, this can be shown through a simple proof. If the two circles have a *common external tangent* of length t, then $t^2 = d^2 - (r_1 - r_2)^2$, so that $t^2/(2r_1 r_2) = 1 - k$. As k is invariant under inversion,

$$\frac{t}{\sqrt{r_1 r_2}}$$

will also be invariant.

Now suppose that four circles with radii r_1, r_2, r_3, and r_4 all touch a circle with radius R externally and that they have common external tangent lines t_{12}, t_{13}, and so on. Invert the figure, choosing the center on the circle with radius R. This is transformed into a line l tangent to the inverted circles with radii r_1', r_2', r_3', and r_4', say at points A_1, A_2, A_3, and A_4, in that order. PTOLEMY's theorem obviously holds for these four collinear points, hence

$$A_1 A_3 \times A_2 A_4 = A_1 A_2 \times A_3 A_4 + A_1 A_4 \times A_2 A_3 \,,$$

but $A_1 A_2 = t_{12}'$, and so on, so that we can also write

$$\frac{t_{13}'}{\sqrt{r_1' r_3'}} \times \frac{t_{24}'}{\sqrt{r_2' r_4'}} = \frac{t_{12}'}{\sqrt{r_1' r_2'}} \times \frac{t_{34}'}{\sqrt{r_3' r_4'}} + \frac{t_{14}'}{\sqrt{r_1' r_4'}} \times \frac{t_{23}'}{\sqrt{r_2' r_3'}} \,.$$

As $t_{12}'/\sqrt{r_1' r_2'} = t_{12}/\sqrt{r_1 r_2}$, and so on, we finally obtain: *if four circles with common external tangent lines t_{12}, and so on, are all tangent, externally, to the same circle, then $t_{13} t_{24} = t_{12} t_{34} + t_{14} t_{23}$.*

This theorem, by CASEY (1866) [Cas2], is a generalization of that of PTOLEMY, which corresponds to the case where the four "circles" all have radius *zero*.

We can show the theorem as follows without inversion (ZACHARIAS, 1942 [Zac1]). Let (Fig. 22.1) M, M_1, and M_2 be the centers of the circles with

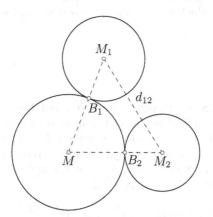

Fig. 22.1.

radii R, r_1, and r_2, $M_1 M_2 = d_{12}$, $\angle M_1 M M_2 = \varphi$, and let B_1 and B_2 be the tangent points, then

$$t_{12}^2 = d_{12}^2 - (r_1 - r_2)^2 \,,$$
$$d_{12}^2 = (R + r_1)^2 + (R + r_2)^2 - 2(R + r_1)(R + r_2) \cos \varphi \,,$$

and

$$(B_1 B_2)^2 = 2R^2 (1 - \cos \varphi) \,,$$

whence it follows that

$$t_{1\,2}^2 = (B_1 B_2)^2 \frac{1}{R^2} (R + r_1)(R + r_2) \,.$$

The theorem now immediately follows by applying PTOLEMY's theorem to the quadrilateral $B_1 B_2 B_3 B_4$.

22.4

CASEY's theorem also holds if one or more circles with radii r_i touch the circle with radius R internally. As is clear by considering the first proof, whenever we have two circles that are tangent in different manners, we must take the common *internal* tangent line t'_{ij} instead of t_{ij}.

   If t, respectively t', does not exist, then we take the formal definition as above, $t^2 = d^2 - (r_1 - r_2)^2$, where t is now imaginary. We see that giving the theorem in a general setting leads to rather complicated choices regarding signs. In fact, we should not state the theorem for four circles, but for four *oriented circles*, to which the clockwise or counterclockwise orientation has been attributed.

22.5

The converse of CASEY's theorem also holds, but the statement and proof encounter analogous difficulties. An application already given by CASEY is to the four circles that are tangent to all (extended) sides of a triangle: if the circle with radius r_4 is the incircle, and those with radii r_1, r_2, and r_3 are the excircles tangent to a, b, and c, then $t_{1\,2} = a + b$, and so on, and $t'_{3\,4} = a - b$, and so on. We conclude from the relation

$$t_{1\,2} t'_{3\,4} + t_{2\,3} t'_{1\,4} + t_{3\,1} t'_{2\,4} = (a^2 - b^2) + (b^2 - c^2) + (c^2 - a^2) = 0$$

that there exists a circle touching the circle with radius r_4 internally and those with radii r_1, r_2, and r_3 externally. It is the *nine-point circle*.

   This result was extended by HART to the figure of four circles tangent to three circles [Har].

Pedal Triangles; Brocard Points

23.1

Let us give a second interpretation of the products p_1, p_2, and p_3 introduced in Chapter 22. Let $D_1D_2D_3$ (Fig. 23.1) be the pedal triangle of triangle ABC for the point D. Then AD is the diameter of the circumcircle of triangle AD_2D_3,

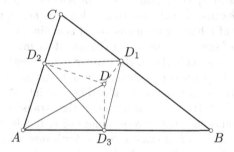

Fig. 23.1.

so that $D_2D_3 = AD\sin A = (AD \times BC)/2R$, where R is the radius of the circumcircle of triangle ABC. It follows that *the products p_1, p_2, and p_3 are in the same proportions as the sides of the pedal triangle of ABC for D*. From this it follows, first of all, that if we consider the four triangles obtained by taking three of the four vertices A,B,C,D, then the pedal triangles of each of these for the remaining vertex are *similar*. Moreover, it turns out that WALLACE's theorem is a direct consequence of PTOLEMY's. Finally, the pedal triangle of a triangle ABC for a point P is similar to the triangle obtained from ABC through an inversion $P(m)$.

O. Bottema, *Topics in Elementary Geometry*,
DOI: 10.1007/978-0-387-78131-0_23, © Springer Science+Business Media, LLC 2008

23.2

The theorem above is useful for determining points whose pedal triangles have given properties. Let us first consider points I for which the pedal triangle of ABC is *equilateral*. If $IA = q_1$, $IB = q_2$, and $IC = q_3$, we must have $aq_1 = bq_2 = cq_3$; in particular, for example, $q_1 : q_2 = b : a$.

The locus of the points whose distances from two fixed points A and B are in a constant ratio k is the perpendicular bisector of AB if $k = 1$, and a circle, called the APOLLONIUS (ca. 250 BC) *circle* (A, B, k), if $k \neq 1$ [Apo], [Hog]. Indeed, if Q is a point of the locus, and S_1 and S_2 are the intersection points of both bisectors of angle AQB and AB, then S_1 and S_2 are fixed points (because $AS : BS = k$), and angle S_1QS_2 is a right angle. Our point I must therefore lie on $(A, B, a/b)$ and $(B, C, c/b)$. If, for example, $a > b$, then A lies inside the first circle and on the second one. The circles therefore have two distinct intersection points I_1 and I_2. The third circle also passes through these points. We have already come across the points I_1 and I_2 in Section 8.3. They are the *isodynamic points* of the triangle. To the property deduced in that chapter, i.e., the distances from I_k, $k = 1, 2$, to the vertices of the triangle are inversely proportional to the opposite sides, can now be added that *the pedal triangle for an isodynamic point is equilateral.*

With each of its isodynamic points, a triangle forms a quadrangle for which the products of the opposite sides are equal, that is, a *harmonic* quadrilateral. The inverse figure of a harmonic quadrilateral is again a harmonic quadrilateral, as the ratios of the products do not change. If we invert with a center on the circumcircle c of triangle ABC, the inverse figure consists of three points on a line l. These also have two isodynamic points, because the definition using the APOLLONIUS circles remains valid. These points are apparently each other's reflections in l. However, an inversion that transforms l into c transforms two such points into two inverse points of c. It follows from this that *the isodynamic points of a triangle are each other's inverse with respect to the circumcircle.* One therefore lies inside the circle c, and the other one outside. According to Section 3.7, the equilateral pedal triangles for these points have *opposite orientations.*

23.3

According to Section 23.1, the points I_1 and I_2 of triangle ABC have the property that an inversion $I_i(m)$ transforms the triangle into an equilateral triangle. If we have two triangles $A_1B_1C_1$ and $A_2B_2C_2$, $A_1'B_1'C_1'$ is an equilateral triangle that is inverse to $A_1B_1C_1$, and $A_2'B_2'C_2'$ is an equilateral triangle that is inverse to $A_2B_2C_2$, then there exists a similarity that transforms $A_1'B_1'C_1'$ into $A_2'B_2'C_2'$. It follows that there always exists a *conformal* transformation that maps $A_1B_1C_1$ into $A_2B_2C_2$. This proves the assertion made in Section 21.4.

If $P_1P_2P_3$ is a given triangle, there always exists a quadrilateral for which the products p_1, p_2, and p_3 are proportional to the sides of $P_1P_2P_3$. Indeed, if O is an arbitrary point and P_1', P_2', and P_3' are the inverse points of P_1, P_2, and P_3 for an inversion $O(m)$, then $OP_1'P_2'P_3'$ is such a quadrilateral. By the above, we can therefore say that if A, B, and C are given points, there always exists a point D such that $BC \times AD$, $CA \times BD$, and $AB \times CD$ are in the same proportions as p_1, p_2, and p_3. That is, there exists a point D for which the pedal triangle of ABC is similar to $P_1P_2P_3$. To find D, we must intersect the two APOLLONIUS circles $(A, B, bp_1/ap_2)$ and $(B, C, cp_2/bp_3)$.

As in Section 23.2, we therefore obtain the following extension: *there always exist two points S_1 and S_2 for which the pedal triangle $A'B'C'$ of a given triangle ABC is similar to a fixed triangle $P_1P_2P_3$. The points S_1 and S_2 are each other's inverses with respect to the circumcircle of ABC. One pedal triangle is directly similar to $P_1P_2P_3$, the other one inversely.*

23.4

The point for which the pedal triangle is directly similar to ABC turns out to be the center of the circumcircle. The points O_1 and O_2 for which the pedal triangles are directly similar to BCA and CAB, respectively, are called BROCARD (1875) points, after their discoverer [Bro]. If q_1, q_2, and q_3 are the distances from O_1 to A, B, and C, then

$$q_1 : q_2 : q_3 = \frac{b}{a} : \frac{c}{b} : \frac{a}{c}.$$

The points O_1 and O_2 can each be constructed by intersecting two APOLLONIUS circles.

Let (Fig. 23.2) $A'B'C'$ be the pedal triangle for O_1; then $\angle A' = \angle B$. As $\angle O_1A'C' = \angle O_1BC'$, we find $\angle O_1BC = \angle O_1A'B'$. The latter, however, is

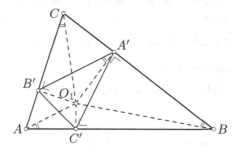

Fig. 23.2.

equal to $\angle O_1CB'$ whence we finally obtain

$$\angle O_1 AB = \angle O_1 BC = \angle O_1 CA \ :$$

BROCARD *angle*, denoted by ω.

If we consider the circle through A, O_1, and B, the inscribed angle BAO_1 intercepts arc $O_1 B$, and therefore so does the equally sized angle $O_1 BC$. In other words, BC touches this circle. We can therefore also determine O_1 as an intersection point of three circles: the one through A that touches BC at B, the one through B that touches CA at C, and the one through C that touches AB at A. The point O_1 consequently lies on the same side of BC as A, and analogously for CA and AB. Hence, O_1, and also O_2, lies inside the triangle. Moreover, as $\angle O_1 A'B' = \omega$, it immediately follows that O_1 is also the first BROCARD point of triangle $A'B'C'$, so that O_1 is the *similarity point* of both similar figures: we can make $A'B'C'$ coincide with BCA by rotating it around O_1 and then applying a central dilation with center O_1 and scaling factor equal to

$$\frac{O_1 B}{O_1 A'} = \frac{1}{\sin \omega} \ .$$

It moreover follows from the similarity of the triangles $O_1 AC'$, $O_1 BA'$, and $O_1 CB'$ that

$$O_1 A' : O_1 B' : O_1 C' = \frac{c}{b} : \frac{a}{c} : \frac{b}{a} \ .$$

By considering the triangles $O_1 AC'$ and $O_1 AB'$, we find

$$\frac{\sin(A - \omega)}{\sin \omega} = \frac{O_1 B'}{O_1 C'} = \frac{a^2}{bc} = \frac{\sin A \sin(B + C)}{\sin B \sin C} \ .$$

After expansion and division by $\sin A$, this gives

$$\cot \omega = \cot A + \cot B + \cot C \ .$$

The symmetry of the outcome shows that the same angle ω is associated to O_2. The scaling factor $1/\sin \omega$ is therefore the same for the pedal triangle for O_2 as for the pedal triangle for O_1. Consequently, the pedal triangles for O_1 and O_2 are congruent. It then follows from Section 3.7 that O_1 and O_2 are equidistant from the center of the circumcircle. As the relation

$$\cot B \cot C + \cot C \cot A + \cot A \cot B = 1$$

holds in every triangle, we have

$$\cot^2 \omega = \cot^2 A + \cot^2 B + \cot^2 C + 2 \ ,$$

and therefore

$$(\cot B - \cot C)^2 + (\cot C - \cot A)^2 + (\cot A - \cot B)^2 = 2(\cot^2 \omega - 3) \ .$$

It follows that $\cot \omega \geq \sqrt{3}$, hence $\omega \leq 30°$, where equality holds only in equilateral triangles.

Chapter 24 contains another property of BROCARD points.

Isogonal Conjugation; the Symmedian Point

24.1

Let ABC be a triangle and let P be a point. In Fig. 24.1, the line CP is

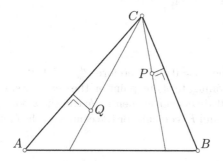

Fig. 24.1.

reflected in the angle bisector of C, taking CP into CQ, where $\angle ACQ = \angle BCP$. Let $\bar{x}$ and $\bar{y}$ be the trilinear coordinates of P and $\bar{x}'$ and $\bar{y}'$ those of Q. Two pairs of similar triangles arise, from which it follows that

$$\bar{x} : \bar{y} = \bar{y}' : \bar{x}', \qquad \text{that is,} \qquad x' : y' = \frac{1}{x} : \frac{1}{y}.$$

The lines CP and CQ are called *isogonal conjugates* with respect to CA and CB. In Fig. 24.2, P and P' are such that CP and CP' are isogonal conjugates with respect to CA and CB and also with respect to BC and BA. Consequently,

$$x' : y' : z' = \frac{1}{x} : \frac{1}{y} : \frac{1}{z},$$

whence it follows that AP and AP' are isogonal conjugates with respect to AB and AC. The conclusion is that for an arbitrary point P, the cevians that

O. Bottema, *Topics in Elementary Geometry*,
DOI: 10.1007/978-0-387-78131-0_24, © Springer Science+Business Media, LLC 2008

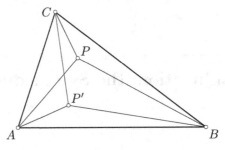

Fig. 24.2.

are the isogonal conjugates of AP, BP, and CP concur. The point P' through which they pass is called the *isogonal conjugate* of P.

The analytic equivalent of this geometrically defined conjugation is

$$x' : y' : z' = yz : zx : xy \ . \tag{24.1}$$

The association of P' to P is therefore *of period two*: if P' is the conjugate of P, then P is the conjugate of P'.

24.2

Let P lie on BC, then $x = 0$ and therefore $P' = (1, 0, 0)$: a vertex of the triangle is the isogonal conjugate of any point on the opposite side. The relationship therefore has so-called *singular* elements. We can show both geometrically and analytically that P and P' coincide for the centers of the four tritangent circles of ABC.

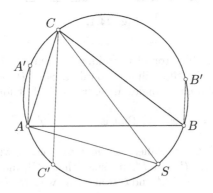

Fig. 24.3.

Let us close the plane for isogonal conjugation by adding the line at infinity. Let AA', BB', and CC' be three parallel lines, and let AS be the isogonal conjugate of AA' (Fig. 24.3), then $\angle CAA' = \angle BAS$, hence arc $CA' =$ arc SB. Moreover arc $CA' =$ arc AC', hence arc $AC' =$ arc SB, from which it follows that CC' and CS are isogonal conjugates. The three cevians isogonal to AA', BB', and CC' therefore concur at S. We conclude that *the conjugate of the line at infinity is the circumcircle*. As in trilinear coordinates, $ax + by + cz = 0$ is the equation of the line at infinity, (24.1) implies that the circumcircle is given by $ayz + bzx + cxy = 0$, a much simpler equation than that of the incircle (Chapter 17, (17.3)). To an arbitrary line $px + qy + rz = 0$, this conjugation associates $pyz + qzx + rxy = 0$, that is, a conic section passing through the vertices A, B, and C. Let (Fig. 24.4) P and P' be isogonal points,

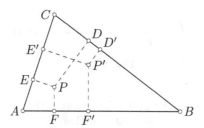

Fig. 24.4.

D, E, and F the projections of P on BC, CA, and AB, and D', E', and F' those of P', then $PE : PF = P'F' : P'E'$. It follows from the similarity of PEA and $P'F'A$ and of PFA and $P'E'A$ that $PE \times AF' = P'F' \times AE$ and $P'E' \times AF = PF \times AE'$, so that $AE' \times AE = AF' \times AF$. That is, E, E', F, and F' lie on a circle c_1, and similarly, F, F', D, and D' lie on a circle c_2, and D, D', E, and E' on a circle c_3. It follows from an argument similar to that of Chapter 5 that the three circles coincide. Therefore, *the pedal triangles of two isogonal points have the same circumcircle*.

24.3

We can easily show that the orthocenter H of a triangle and the center O of the circumcircle are isogonal conjugates, and that the two BROCARD points are each other's conjugates. Isogonal conjugation therefore does not introduce new special points as conjugates of H and O. This is not the case for the centroid G.

The isogonal conjugate of G is a new special point, often denoted by K and called the symmedian point [Hon].

The barycentric coordinates of G are all equal. Consequently, the trilinear coordinates are proportional to a^{-1}, b^{-1}, and c^{-1}. It therefore follows from (24.1) that *the distances from K to the sides are proportional to those* *sides:* $K = (a, b, c)$. The cevian isogonal to a triangle median is called a *sym-median*. The point K is therefore the intersection point of the three symmedians. The barycentric coordinates of K are proportional to a^2, b^2, and c^2. It follows that *a symmedian divides the opposite side into segments proportional to the squares of the adjacent sides.* Let (Fig. 24.5) $A'B'C'$ be the pedal tri-

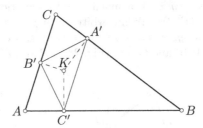

Fig. 24.5.

angle for K. The distances from K to the sides satisfy $KA' = pa$, $KB' = pb$, and $KC' = pc$, where p is a constant. We then have

$$[KB'C'] = \tfrac{1}{2}KB' \times KC' \sin \alpha = \tfrac{1}{2}pb \times pc \times \sin \alpha = p^2[ABC] \ .$$

It follows that the areas $[KB'C']$, $[KC'A']$, and $[KA'B']$ are equal; that is, K *is the centroid of its own pedal triangle.*

According to LAGRANGE's *identity*, of which we have already encountered the special case $x = y = z = 1$ in Section 11.5, we have

$$(x^2 + y^2 + z^2)(a^2 + b^2 + c^2) = $$
$$(ax + by + cz)^2 + (cy - bz)^2 + (cx - az)^2 + (ay - bx)^2 \ .$$

Let a, b, and c be the sides of a triangle and x, y, and z the distances from a point to those sides, then $ax + by + cz = 2[ABC]$, whence it follows that

$$x^2 + y^2 + z^2 \geq \frac{4[ABC]^2}{a^2 + b^2 + c^2} \ ,$$

where equality holds only if $x : y : z = a : b : c$. The conclusion is that *the point K is the point for which the sum of the squares of the distances from the sides is minimal.*

Let AC' and BC' be the tangents to the circumcircle of ABC at A and B (Fig. 24.6). Then C' is the pole of AB. Let C_1 and C_2 be the projections from C' on CB and CA. Then $\angle C'AC_2$ is a secant-tangent angle that intercepts arc AC, and is therefore equal to β. Likewise, $\angle C'BC_1$ is equal to α. As $AC' = BC'$, we have $C'C_1 : C'C_2 = \sin \alpha : \sin \beta = a : b$, or in other words, CC' is the symmedian of C. It follows that K *is the intersection point*

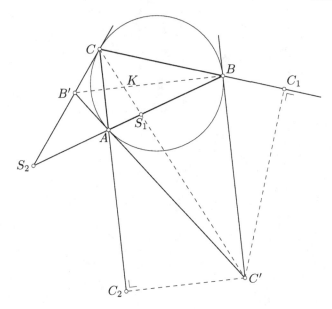

Fig. 24.6.

of three lines, each of which joins one of the vertices of ABC and the pole of the opposite side with respect to the circumcircle.

We conclude our thoughts on the symmedian point with the following statement: *K is the intersection point of three lines, each of which joins the midpoint of a side with the midpoint of the altitude on that side.*

Let us give an analytic proof using trilinear coordinates. Let S_1 be the midpoint of BC and S_2 that of the altitude from A. Then $S_1 = (0, c, b)$ and $S_2 = (1, \cos\gamma, \cos\beta)$, and the line joining them is given by

$$(b\cos\gamma - c\cos\beta)x - by + cz = 0,$$

that is,

$$(b^2 - c^2)x - aby + acz = 0.$$

The point $K = (a, b, c)$ satisfies this equation. Likewise, it satisfies the equations of the two analogous joins.

25

Isotomic Conjugation

In Fig. 25.1, triangle ABC is given, as is a point P. Let S be the intersection point of CP and AB. We reflect S in the midpoint of AB, giving the point S'; in other words, $AS' = BS$. The points S and S' divide AB into the same two segments, but in inverse order. We carry out the analogous construction

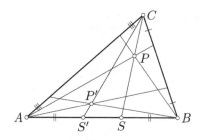

Fig. 25.1.

on AC and BC. CEVA's theorem provides a direct proof for the statement that the cevians through the three new points concur at a point P'. The relationship between P and P' is called *isotomic conjugation*. It resembles isogonal conjugation, but now concerns equal line segments instead of equal angles.

The new relationship comes from affine geometry and barycentric coordinates seem a logical choice for an analytic study. Let $P = (X, Y, Z)$, then $P' = (X^{-1}, Y^{-1}, Z^{-1})$, or, analogously to Section 24.1,

$$X' : Y' : Z' = YZ : ZX : XY . \tag{25.1}$$

Isotomic conjugation is also of period two and is an involution. To a vertex it associates every point of the opposite side. The centroid $(1, 1, 1)$ is invariant, as are the points $(-1, 1, 1)$, $(1, -1, 1)$, and $(1, 1, -1)$, which are the vertices

O. Bottema, *Topics in Elementary Geometry*,
DOI: 10.1007/978-0-387-78131-0_25, © Springer Science+Business Media, LLC 2008

of the circumscribed homothetic triangle. The GERGONNE and NAGEL points studied in Section 2.6 are isotomic conjugates.

To the line at infinity with equation $X + Y + Z = 0$ is associated the figure with equation $YZ + ZX + XY = 0$. This is not the circumcircle, a notion that is unknown in affine geometry, but a well-determined ellipse that goes through the vertices A, B, and C and at those points touches the line parallel to the opposite side. The center of this circumellipse that is named after STEINER is the centroid of the triangle [Ste], [Wel]. Unlike isogonal conjugation, which gave us the symmedian point with its rich collection of properties, isotomic conjugation does not seem to give rise to any new special point.

Triangles with Two Equal Angle Bisectors

26.1

The fact that a triangle with two equal altitudes is isosceles follows directly from the congruence of two right triangles inside the original triangle. The same conclusion holds when two medians are equal; a proof can be given using the formula for the length of a median as a function of the sides.

The statement that a triangle with two equal (interior) angle bisectors is isosceles will surprise no one, but strangely enough the proof is not simple at all. Approximately a century and a half ago, the unknown mathematician LEHMUS asked STEINER for a proof and the latter published a geometric one (1844) [Lan], [Ste]. The resulting theorem is often called the STEINER-LEHMUS theorem.

Last century a number of proofs were given, of which we include three here.

26.2

The idea of the algebraic proof that follows is clear: express the lengths of the angle bisectors AD and BE of triangle ABC in terms of the sides and let the answers agree. The formula $AD^2 = bc - pq$, with $p = CD$ and $q = BD$, is well known. As $p : q = c : b$, this gives

$$AD^2 = \frac{bc(a + b + c)(-a + b + c)}{(b + c)^2} \,,$$

and analogously

$$BE^2 = \frac{ac(a + b + c)(a - b + c)}{(a + c)^2} \,.$$

The equality $AD^2 = BE^2$ is of course satisfied for $a = b$ (in an isosceles triangle there are two equal angle bisectors) and after some computation we find

O. Bottema, *Topics in Elementary Geometry*,
DOI: 10.1007/978-0-387-78131-0_26, © Springer Science+Business Media, LLC 2008

$$(a - b)\big(c^3 + (a + b)c^2 + 3abc + ab(a + b)\big) = 0 \ . \qquad (26.1)$$

As the second factor on the left consists of strictly positive terms, the conclusion is that $a = b$.

26.3

Here follows a short trigonometric proof, by contradiction. We have

$$AD \times (b + c) \sin \tfrac{1}{2}\alpha = 2[ABC] \ \text{and} \ BE \times (a + c) \sin \tfrac{1}{2}\beta = 2[ABC] \ ,$$

so that $AD = BE$ gives

$$b \sin \tfrac{1}{2}\alpha - a \sin \tfrac{1}{2}\beta = c(\sin \tfrac{1}{2}\beta - \sin \tfrac{1}{2}\alpha) \ . \qquad (26.2)$$

If $\beta > \alpha$, hence $\cos(\alpha/2) > \cos(\beta/2)$, then it follows from $b \sin \alpha = a \sin \beta$, that is, $b \sin(\alpha/2) \cos(\alpha/2) - a \sin(\beta/2) \cos(\beta/2) = 0$, that the left-hand side of (26.2) is negative, while the right-hand side is positive. As the assumption $\beta < \alpha$ also leads to a contradiction, $\beta = \alpha$.

26.4

A more geometric proof (DESCUBE, 1880 [FGM]), also by contradiction, goes as follows (Fig. 26.1). Consider the parallelogram $ADFE$ and draw BF. Then

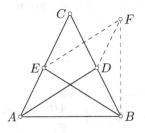

Fig. 26.1.

$EF = AD = BE$, hence $\angle EFB = \angle EBF$. If $a > b$, then $\angle EFD > \angle EBD$, hence $\angle DFB < \angle DBF$, that is, $DB < AE$. It then follows from the triangles ABD and BAE that $a < b$. The assumption therefore leads to a contradiction. As the same holds for $a < b$, we have $a = b$.

26.5

A triangle with two equal *exterior* angle bisectors does not have to be isosceles. Instead of relation (26.1), the computation now gives

$$(a - b)\big(c^3 - (a + b)c^2 + 3abc - ab(a + b)\big) = 0 \, . \qquad (26.3)$$

We can show that there exist positive numbers a, b, and c satisfying the triangular inequalities for which the second factor on the left is zero. The example $\alpha = 12°$, $\beta = 132°$, and $\gamma = 36°$ with $AD' = BE'$ found by EMMERICH is well known; see Fig. 26.2 [Emm].

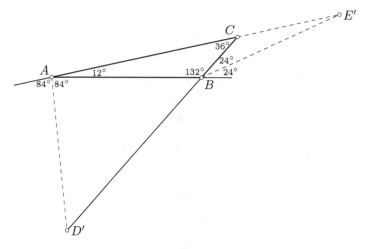

Fig. 26.2.

The Inscribed Triangle with the Smallest Perimeter; the Fermat Point

27.1

Let ABC (Fig. 27.1) be an acute triangle, and let P, Q, and R be points on AB, BC, and CA, in that order. How must we choose these points in order to minimize the *perimeter* of the triangle PQR inscribed in triangle ABC? Let us begin by considering triangles PQR with P fixed. Let P_1 and P_2 be

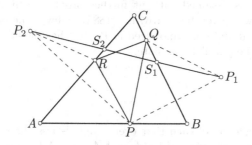

Fig. 27.1.

the reflections of P in the sides BC and AC, then $PQ = P_1Q$ and $PR = P_2R$, so that the perimeter of triangle PQR is equal to the length of the broken line P_1QRP_2. For fixed P, P_1 and P_2 are also fixed. Of all triangles PQR, PS_1S_2 therefore has the smallest perimeter, where S_1 and S_2 are the intersection points of P_1P_2 with CB and CA. This perimeter is equal to P_1P_2.

Next let P be arbitrary and let us determine the minimum of P_1P_2. Even if the points P_1 and P_2 are no longer fixed, the triangle CP_1P_2 still has a number of fixed properties. Indeed, CP_1 and CP_2 are both equal to CP and the said triangle is therefore isosceles. Moreover, angle P_1CB is equal to PCB and P_2CA to PCA, so that angle P_1CP_2 is twice angle C. As C is acute, P_1CP_2 will remain smaller than a straight angle. It follows that P_1P_2 always intersects the sides CB and CA, and not their extensions, and that to each choice of P

O. Bottema, *Topics in Elementary Geometry*,
DOI: 10.1007/978-0-387-78131-0_27, © Springer Science+Business Media, LLC 2008

on AB corresponds a pair of points S_1 and S_2. Triangle P_1CP_2 therefore has a constant top angle, and the basis is minimal if the leg is minimal, hence if CP is minimal. Because of the acuteness of A and B, P must consequently be chosen at the foot of the altitude from C. The points S_1 and S_2 will lie at the feet of the altitudes from A and B. We therefore have that *of all triangles inscribed in an acute triangle, the pedal triangle for the orthocenter, also called the orthic triangle, has the smallest perimeter.*

If angle A, for example, is right or obtuse, then CA is the shortest of all lines CP, which implies that in a right or obtuse triangle twice the shortest altitude is the "triangle" with the smallest perimeter.

The theorem has been known for a long time. If $A'B'C'$ is the orthic triangle, the angles $C'A'B$ and $B'A'C$ are equal, as are $A'B'C$ and $C'B'A$ and $B'C'A$ and $A'C'B$. A ray of light can therefore be run along the perimeter of triangle $A'B'C''$, reflecting in the sides of triangle ABC. Let Q and R be two points on the same side of the line BC, and P an arbitrary point on this line, then the shortest broken line QPR is the one where the angles QPB and RPC are equal. If PQR is inscribed in ABC and does not coincide with the orthic triangle, then by fixing two vertices and changing the third one appropriately, we can obtain an inscribed triangle with smaller perimeter. The proof based on this idea seems to have the same type of flaws as the proofs of STEINER (Section 13.4) for the isoperimetric inequality, because here too the existence of the minimum is assumed without further proof. The first correct proof of the theorem was given by SCHWARZ (1884) [Schw]. The one given above comes from FEJÉR and was announced in 1930. It has the advantage of being independent of the parallel postulate.

27.2

Consider a triangle ABC such that every angle is less than $120°$. Choose a point X inside the triangle and let $d_1 = XA$, $d_2 = XB$, and $d_3 = XC$. Now consider the following question, first asked by FERMAT [Fer] and answered by TORRICELLI (1659): for what point X is $d_1 + d_2 + d_3$ minimal [Viv]?

The following proof was given by HOFMANN (1929) [Hof]. Consider the equilateral triangle CXX' with side d_3, drawn outward on the side CX of triangle CXA (Fig. 27.2). Next, the triangle $CX'A'$, congruent to triangle CXB, is drawn on side CX' of triangle CXX', again outward. We then have $AX = d_1$, $XX' = d_3$, and $X'A' = d_2$, which implies that the length of the broken line $AXX'A'$ is equal to $d_1 + d_2 + d_3$, whose minimum we wish to determine.

It turns out that the point A' is independent of X. Indeed, $CA' = a$ and $\angle BCA' = 60°$, because $\angle X'CA' = \angle XCB$. The point A' is therefore the top of the equilateral triangle drawn outward on BC. The desired point, called the FERMAT point (sometimes also called the TORRICELLI point, or the FERMAT-TORRICELLI point), and denoted by F, is therefore the intersection

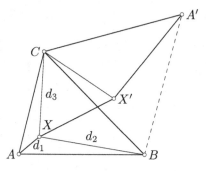

Fig. 27.2.

point of AA' with the analogous lines BB' and CC', whose concurrence has already been shown in Chapter 11.

If triangle ABC has an angle that is greater than or equal to 120°, then F lies at the vertex of the obtuse angle.

27.3

The point F also has the following three properties, of which we leave the proofs to the reader.

I. If $FA + FB + FC = S$, then $S^2 = \frac{1}{2}(a^2 + b^2 + c^2) + 2\sqrt{3} \times O$. It follows from the symmetry of the outcome that $AA' = BB' = CC'$.

II. The point F is the *first isogonic center* of the triangle, meaning that $\angle BFC = \angle CFA = \angle AFB \ (= 120°)$.

III. The point F is the isogonal conjugate of one of the isodynamic points.

Appendix: Remarks and Hints by J.M. Aarts

Much has changed in the teaching of mathematics since the first edition of this book. Nowadays much less time is spent on Euclidean geometry than before. Because of this a present-day reader might have some difficulties when studying this book. The author uses notions and results that were once common knowledge, but which these days are known only to the older generation.

This appendix contains a number of hints to help the reader understand the text.

Chapter 1

In Section 1.5, the areas of the triangles 1, 2, 3, and 4 used in the dissection proof of the Pythagorean theorem are mentioned. The formulas given there can be deduced as follows. In Fig. 1.4, the bisector CD of angle ACB divides the triangle ABC into pieces 2 and 3. This implies that their areas are in the ratio $a : b$. Moreover, we easily see that pieces 1 and 3 are similar. Using the remark in Section 1.2, we find that the areas of these are in the ratio $a^2 : b^2$. In the same way, we determine that the areas of pieces 2 and 4 are in the ratio $a^2 : b^2$.

Chapter 2

In Section 2.2, we encounter the notion of *locus*. Instead of *the segment d_a is the locus of the points equidistant from AB and AC*, we would now say *the segment d_a is the set of all points equidistant from AB and AC*.

At the end of Section 2.6, the formula $BA' = s - c$ is mentioned. Here s is the semiperimeter of triangle ABC. In Fig. 2.5, we find $BC' = BA'$, $CA' = CB'$, and $AC' = AB' = s$; hence $BA' = s - c$.

Chapter 3

In Section 3.1, the second proof of the concurrence of the altitudes is based on the fact that the triangles $AB'C'$ and so on are similar to ABC (Fig. 3.2). To prove this, we first show that $AA'B$ is similar to triangle $CC'B$. It follows that

$AB : CB = BA' : BC'$. Using this, we show that triangle ABC is similar to $A'BC'$. In particular, $\angle BA'C' = \angle CAB$ and $\angle A'C'B = \angle BCA$. Because of this, we call $A'C'$ and CA *antiparallel* with respect to the sides of angle B. See also Chapter 5.

In Section 3.7, the following remarks may be useful in determining the formula for the area $[A'B'C']$ of the pedal triangle $A'B'C'$. First of all, we show that $B'C' = TA \times \sin A$ (and then analogously $A'C' = TB \times \sin B$). This can be deduced as follows using the law of sines (see the remark in Chapter 5). We refer to Fig. 3.7. In triangle $C'B'A$, we have

$$\frac{B'C'}{\sin A} = \frac{AB'}{\sin \angle AC'B} \, .$$

As $AB'TC'$ is a cyclic quadrilateral, we have $\angle AC'B' = \angle ATB'$. Therefore

$$\frac{B'C'}{\sin A} = \frac{AB'}{\sin \angle AC'B} = AT \, ,$$

from which the formula above follows. Moreover, $TB \times \sin \angle TBS = TS \times \sin C$ because both sides are equal to the length of the perpendicular from T on BS.

Chapter 5

The deduction uses the formulas $a = 2R \sin \alpha$, and so on, without mentioning them explicitly. The *law of sines*

$$\frac{a}{\sin \alpha} = \frac{b}{\sin \beta} = \frac{c}{\sin \gamma}$$

and the formula $[ABC] = abc/4R$ for the area of triangle ABC can easily be deduced from these. See Fig. 6.2.

Chapter 6

Formula (6.3) says that there exists a $k \neq 0$ such that

$$kx = (1 - \lambda)x_1 + \lambda x_2 \, ,$$
$$ky = (1 - \lambda)y_1 + \lambda y_2 \, ,$$
$$kz = (1 - \lambda)z_1 + \lambda z_2 \, .$$

These are three equalities in two unknowns k and λ. This system has a nontrivial solution only if the equations are independent. This leads to Condition (6.5), or to the equivalent Condition (6.4).

The following should help clarify Formula (6.9) and its deduction. The equation for line OH is

$$\begin{vmatrix} x & y & z \\ \cos \alpha & \cos \beta & \cos \gamma \\ \cos \beta \cos \gamma & \cos \gamma \cos \alpha & \cos \alpha \cos \beta \end{vmatrix} = 0 \, .$$

That is,

$$x \cos \alpha (\cos^2 \beta - \cos^2 \gamma) + y \cos \beta (\cos^2 \gamma - \cos^2 \alpha) + z \cos \gamma (\cos^2 \alpha - \cos^2 \beta) = 0.$$

The last equation can also be written more succinctly as

$$\sum x \cos \alpha \left(\cos^2 \beta - \cos^2 \gamma \right) = 0 \,.$$

The summation sign is an abbreviation for the summation over all cyclic permutations of x, y, z, of α, β, γ, of a, b, c, and so on, depending on the context. To obtain Formula (6.9) from this one, we use the law of cosines $a^2 = b^2 + c^2 - 2bc \cos \alpha$. This gives

$$\sum x \cos \alpha (\cos^2 \beta - \cos^2 \gamma) = \frac{1}{2abc} \sum ax \left(b^2 + c^2 - a^2 \right) (\sin^2 \gamma - \sin^2 \beta)$$

$$= \frac{-1}{8abcR^2} \sum ax (b^2 + c^2 - a^2)(b^2 - c^2)$$

$$= 0 \,.$$

Chapter 7

We once more use the summation convention of the previous chapter. In the deduction of Formulas (7.5) and (7.7), the following formula for the area of triangle ABC (*Heron's formula*) is used:

$$[ABC]^2 = s(s-a)(s-b)(s-c) = \tfrac{1}{16} \left(2 \sum a^2 b^2 - \sum a^4 \right) ,$$

where s is the semiperimeter of triangle ABC.

Chapter 8

Here is a hint for the deduction of the trigonometric formula (8.1):

$$\cos^2 \varphi_1 + \cos^2 \varphi_2 + \cos^2 \varphi_3 = \cos^2 \varphi_1 + \sin^2(90^\circ - \varphi_2) - \sin^2 \varphi_3 + 1$$

$$= \cos^2 \varphi_1 + \sin(90^\circ - \varphi_2 + \varphi_3) \cos \varphi_1 + 1$$

$$= 2 \cos \varphi_1 \cos \varphi_2 \cos \varphi_3 + 1 \,.$$

In Section 8.2, the law of cosines is used for (8.6). In Section 8.3, (8.13) is obtained by eliminating λ_1, λ_2, and λ_3 in (8.9), (8.10), and (8.12).

Chapter 10

In the computation of EF using the law of cosines, the following trigonometric formula plays a role: if $\alpha + \beta + \gamma = 180^\circ$, then

$$\sin^2 \alpha + \sin^2 \beta + \cos^2 \gamma = 1 + 2 \sin \alpha \sin \beta \cos \gamma \,.$$

Chapter 11

The formula for HO^2 in Section 11.4 can be found as follows using results from Chapter 6. In Figs. 6.2 and 6.3, we have $BH_1 = c \cos \beta$, $BM_1 = a/2$, $HH_1 = 2R \cos \beta \cos \gamma$, and $OM_1 = R \cos \alpha$. Consequently,

$$\begin{aligned} OH^2 &= H_1 M_1^2 + (HH_1 - OM_1)^2 \\ &= (\tfrac{1}{2}a - c \cos \beta)^2 + (2R \cos \beta \cos \gamma - R \cos \alpha)^2 \\ &= R^2 - ac \cos \beta + c^2 \cos^2 \beta + 4R^2 \cos^2 \beta \cos^2 \gamma - 4R^2 \cos \alpha \cos \beta \cos \gamma \ . \end{aligned}$$

The desired expression for HO^2 now follows using

$$c^2 \cos^2 \beta - ac \cos \beta + 4R^2 \cos^2 \beta \cos^2 \gamma = -4R^2 \cos \alpha \cos \beta \cos \gamma \ ,$$

an expression that in turn can be proved using the law of sines.

Chapter 13

In Section 13.2, the second ratio $AD : DS : SA$ can be deduced from the first one by solving for x/d and y/d.

Chapter 16

For Formula (16.3) see the remark for Chapter 8. Note that $\alpha + \beta + \gamma = 180°$. The formula for HM^2 in Section 16.3 is deduced in the remark for Chapter 11.

Chapter 17

In Section 17.1, the trilinear coordinates of the *center of mass G of the perimeter of triangle ABC* are mentioned. We might deduce the formulas as follows. Imagine the perimeter of triangle ABC as formed by a homogeneous thread. The perimeter consists of three pieces with respective lengths a, b, and c. We replace these pieces by masses a, b, and c placed at the midpoints P, Q, and R of respectively BC, CA, and AB. We replace the masses at Q and R by one mass $b + c$ at the point S on RQ chosen such that $c \times RS = b \times SQ$. The moments of the mass c at R and of the mass b at Q with respect to S are then equal, but opposite. The point G that we are looking for lies on the join SP. Through a cyclic permutation of A, B, and C, and so on, we find that G coincides with the center of the incircle of triangle PQR. Let us compute the trilinear coordinates of a number of points: $A = (h_a, 0, 0)$, where h_a is the altitude in A; $P = (0, h_b, h_c)$, $Q = (h_a, 0, h_c)$, and $R = (h_a, h_b, 0)$. It follows that $S = (h_a, ch_b/(b + c), bh_c/(b + c))$. The line SP has equation

$$h_b h_c \left(\tfrac{c}{b+c} - \tfrac{b}{b+c} \right) x - h_a h_c y + h_a h_b z = 0 \ .$$

The point G lies on line SP. Through a cyclic permutation of a, b, and c we find two more lines on which G lies. Through substitution we can verify that

$$G = (h_a(b + c), h_b(c + a), h_c(a + b)) = (x, y, z) \ .$$

The trilinear coordinates $\bar{x}$, $\bar{y}$, and $\bar{z}$ are then found through normalization: $\bar{x} = kx$, and so on, where $a\bar{x} + b\bar{y} + c\bar{z} = 2[ABC]$.

Chapter 22

Section 22.2 contains an extension of the law of cosines. That formula can be deduced from the above by applying the usual law of cosines to triangle $A'B'C'$, where A', B', and C' are obtained, in that order, out of A, B, and C through the inversion $D(m)$. We find

$$(A'C')^2 = (A'B')^2 + (B'C')^2 - 2 \times A'B' \times B'C' \cos A'B'C' \ .$$

Now fill in the values found for $A'B'$, and so on.

In Section 22.3, in the first formula, φ is the angle between the rays that join the intersection point of the circles to the centers.

Chapter 23

For the deductions in Section 23.1, we refer to the remarks made for Section 3.7 and Chapter 5.

The formula $\sum \cot B \cot C = 1$ in Section 23.4 can be found simply by expanding $\sin(A + B + C) = 0$.

Additional literature

N. ALTSHILLER-COURT, *College Geometry: an introduction to the geometry of the triangle and the circle*, 2nd ed., Barnes and Noble, New York, 1952.

M. BERGER, *Geometry I*, Springer, Berlin, 1987.

M. BERGER, *Geometry II*, Springer, Berlin, 1987.

H.S.M. COXETER, *Introduction to Geometry*, New York, 1989.

H.S.M. COXETER AND S.L. GREITZER, *Geometry Revisited*, New Mathematical Library 19, Mathematical Association of America, 1967.

R. HONSBERGER, *Episodes in nineteenth and twentieth Century Euclidean Geometry*, New Mathematical Library 37, Mathematical Association of America, 1995.

M. KOECHNER, *Lineare Algebra und analytische Geometrie,*, Springer, Berlin, 1992.

M. KOECHNER AND A. KRIEG, *Ebene Geometrie*, Springer, Berlin, 1993.

Bibliography

[Apo]   Apollonius: Conics (250).

[Bar]   Barrau, J.A.: Bijdragen tot de theorie der configuraties. Ph.D. thesis, Universiteit van Amsterdam (1907).

[Ben]   Benson, D.C.: Sharpened Forms of the Plane Isoperimetric Inequality. Amer. Math. Monthly, **77** (1), 29–34 (Jan. 1970).

[Ber]   Berkhan, G., Meyer, W.Fr.: Neuere Dreiecksgeometrie, Encyclopaedie der Mathematischen Wissenschaften, **III/1** (10). Leipzig (1920).

[Bes]   Besthorn, R.O. et al.: Codex Leidensis 399,1. Euclidis Elementa ex interpretatione al-Hadschdschadschii cum commentariis Al-Narizii, Pars I. Copenhagen (1897).

[Bha]   Bhaskara: Bijaganita (Seed Counting or Root Extraction) (1150).

[BKOR]   Bos, H.J.M., Kers, C., Oort, F., Raven, D.W.: Poncelet's Closure Theorem, Its History, Its Modern Formulation, a Comparison of Its Modern Proof with Those by Poncelet and Jacobi, and Some Mathematical Remarks Inspired by These Early Proofs. Expo. Math., **5**, 289–364 (1987).

[Bla]   Blaschke, W.: Kreis und Kugel. Verlag von Veit & Comp., Leipzig (1916).

[Bol1]   Bol, G.: Zur Theorie der konvexen Körper. Jahresber. Deutsch. Math.-Verein., **49**, 113–123 (1939).

[Bol2]   Bol, G.: Einfache Isoperimetriebeweise für Kreis und Kugel. Abh. Math. Sem. Hansischen Univ., **15**, 27–36 (1943).

[Bos]   Bosse, A.: Manière universelle de Mr Desargues, pour praticquer la Perspective par petit-pied, comme le géométral. Pierre Deshayes, Paris (1648).

[Bot1]   Bottema, O.: De elementaire meetkunde van het platte vlak. Noordhoff, Groningen (1938).

[Bot2]   Bottema, O. et al.: Geometric Inequalities. Wolters-Noordhoff, Groningen (1969).

[Bri]   Brianchon, C.: Mémoire sur les surfaces courbes du second degré. J. Ecole Polytechnique, **13** (1806).

[BP]   Brianchon, C.J., Poncelet, J.V.: Recherches sur la détermination d'une hyperbole équilatère, au moyen de quatres conditions données. J. Ecole Polytechnique (1820).

[Bro]   Brocard, H.: Question 1166. Nouvelles Annales de Mathématiques. Deuxième Série, **14**, 192 (1875).

134   Bibliography

[Can]   Cantor, M.B.: Vorlesungen über Geschichte der Mathematik. 4 Bde.
           Teubner, Leipzig (1908), reprinted (1965).
[Cas1]   Casey, J.: A sequel to the first six books of the Elements of Euclid. Dublin
           (1881).
[Cas2]   Casey, J.: On the Equations and Properties–(1) of the System of Circles
           Touching Three Circles in a Plane; (2) of the System of Spheres Touching
           Four Spheres in Space; (3) of the System of Circles Touching Three Cir-
           cles on a Sphere; (4) of the System of Conics Inscribed to a Conic, and
           Touching Three Inscribed Conics in a Plane. Math. Proc. R. Ir. Acad.,
           9, 396–423 (1864–1866).
[Cay]   Cayley, A.: Philos. Math., 6, 99–102 (1853).
[Cev]   Ceva, G.: De Lineis Rectis se Invicem Secantibus Statica Constructio.
           Milano (1678).
[Cha]   Chapple, W.: An essay on the properties of triangles inscribed in and
           circumscribed about two given circles. Misc. Curr. Math., 4, 117–124
           (1746).
[Col]   Colebrooke, H.T.: Algebra, with Arithmetic and Mensuration from the
           Sanscrit of Brahmagupta and Bhaskarahmagupta and Bhaskara. John
           Murray, London (1817), reprinted by Sharada Publishing House, Delhi
           (2005).
[Dar]   Darboux, G.: Nouvelles Annales de Math., 2 (5) (1866).
[Emm]   Emmerich, A.: Die Brocardschen Gebilde und ihre Beziehungen zu den
           verwandten merkwürdigen Punkten und Kreisen des Dreiecks. Reimer,
           Berlin (1891).
[Eul]   Euler: Solutio Facilis Problematum Quorumdam Geometricorum Dif-
           ficillimorum. Novi Commentarii Academiae Scientiarum imperialis
           Petropolitanae, 11, 103–123 (1765).
[Fer]   Fermat, P.: Methodus ad disquirendam maxima et minimam. Letter to
           Descartes, who received it on January 10, 1638.
[Feu]   Feuerbach, K.W.: Eigenschaften einiger merkwürdigen Punkte des ger-
           adlinigen Dreiecks und mehrerer durch sie bestimmten Linien und Fig-
           uren: Eine analytisch-trigonometrische Abhandlung (Properties of some
           special points in the plane of a triangle, and various lines and figures
           determined by these points: an analytic-trigonometric treatment). Riegel
           und Wiesner, Nürnberg (1822).
[FGM]   F. G.-M. (Frère Gabriel-Marie): Exercices de Géométrie, 6th ed., 1920;
           Gabay reprint, Paris (1991).
[FH]   Finsler, P., Hadwiger, H.: Einige Relationen im Dreieck. Comment. Math.
           Helv., 10, 316–326 (1937).
[Fus]   Fuss, N. von: De quadrilateris, quibus circulum tam inscribere quam
           circumscribere licet. Nova Acta Academiae Scientiarum Imperialis
           Petropolitanae, 10, 103–125 (1792).
[Gal]   Gallatly, W.: The modern geometry of the triangle. Francis Hodgson,
           London (1910).
[Gard]   Gardner, M.: The Pythagorean Theorem. In: Gardner, M.: Sixth Book of
           Mathematical Games from Scientific American. Charles Scribner's Sons,
           New York (1971).
[Garf]   Garfield, J.A.: New Engl. J. Educ., 3, 161 (1876).
[Gau]   Gauss, C.F.: Monatl. Korresp., 22 (1810).
[Gep]   Geppert, H.: Über den Brunn-Minkowskischen Satz. Math. Z., 42, 238–
           254 (1937).

[Gro]   Groenman, J.T.: Problem 1432. Crux Math., **110** (July 1989).
[Har]   Hart, A.: On the Extension of Terquem's Theorem. Q. J. Math., **IV**, 260–261 (1860).
[Hea1]   Heath, T.L.: Euclid: The Thirteen Books of Euclid's Elements, Vol. 1 (2nd ed.). Dover, New York (1956).
[Hea2]   Heath, T.L.: A History of Greek Mathematics. Dover (1981).
[Hes]   Hessenberg, G.: Beweis des Desarguesschen Satzes aus dem Pascalschen. Math. Ann., **61**, 161–172 (1905).
[Hil]   Hilbert, D.: Grundlagen der Geometrie. In: Festschrift zur Feier der Enthüllung des Gauss-Weber Denkmals in Göttingen. Teubner, Leipzig, (1899).
[Hof]   Hofmann, J.E.: Elementare Lösung einer minimumsaufgabe. Math. Z., **60**, 22–23 (1929).
[Hog]   Hogendijk, J.H.: Desargues' 'Brouillon project' and the 'Conics' of Apollonius. Centaurus, **34** (1), 1–43 (1991).
[Hon]   Honsberger, R.: Episodes in Nineteenth and Twentieth Century Euclidean Geometry. Mathematical Association of America, Washington, D.C. (1995).
[IJz]   IJzeren, J. van: De stelling van Morley in verband met een merkwaardig soort zeshoeken. Euclides, **14**, 277–284 (1937).
[Jac]   Jacobi, C.G.J.: Über die Anwendung der elliptischen Transcendenten auf ein bekanntes Problem der Elementargeometrie. J. Reine Angew. Math., **3**, 376–389 (1828).
[Joh1]   Johnson, R.A.: Advanced Euclidean Geometry. Boston (1929), reprinted by Dover Publications, New York (1960).
[Joh2]   Johnson, R.A.: Modern Geometry: An Elementary Treatise on the Geometry of the Triangle and the Circle. Houghton Mifflin, Boston, MA (1929).
[Joh3]   Johnson, R.A.: Discussions: Relating to the "Simson Line" or "Wallace Line". Amer. Math. Monthly, **23** (61) (1916).
[Kub]   Kubota, T.: Einige Ungleichheitsbeziehungen über Eilinien und Eiflächen. Sci. Rep. Tôhuku Imp. Univ., **1** (12), 45–65 (1923).
[Lag]   Lagrange, J.L.: Oeuvres de Lagrange, **4** (1869).
[Lal]   Lalesco, T.: La géométrie du triangle. Vuibert, Paris (1937), reprinted by Jacques Gabay, Paris (1987).
[Lan]   Lange, T.: Nachtrag zu dem Aufsatze in Thl. XIII, Nr. XXXIII. Arch. Math. Phys., **15**, 221–226 (1850).
[Lem]   Lemoine, E.: Géométrographie ou art des constructions géométriques. C. Naud, Paris (1902).
[LH]   Longuet-Higgins, M.S.: A fourfold point of concurrence lying on the Euler line of a triangle. Mathematical Intelligencer, **22**, 54–59 (2000).
[Lio]   Liouville, J.: Note au sujet de l'article précédent. J. Math. Pures Appl., **12**, 265–290 (1847).
[Luc1]   Lucas, F.E.A.: Nouv. Ann. Math., 240 (1876).
[Luc2]   Lucas, F.E.A.: Nouv. Corr. Math. (1876, 1878).
[Mac]   Mackay, J.S.: Historical notes on a geometrical theorem and its developments (18th century). Proc. Edinburgh Math. Soc., **5**, 62–78 (1887).
[Men]   Menelaus of Alexandria: Sphaerica Book 3 (ca. 100).
[Min]   Minkowski, H.: Volumen und Oberfläche. Math. Ann., **57**, 447–495 (1903).

[Möb]   Möbius, A.F.: Der barycentrische Calcul.: ein neues Hülfsmittel zur analytischen Behandlung der Geometrie. Leipzig (1827).

[Mol]   Molenbroek, P.: Leerboek der vlakke meetkunde. Noordhoff, Groningen (1943).

[Mul]   Multatuli: Ideeën, Tweede bundel met een naw. van J.J. Oversteegen. Multatuli/Querido (1985).

[Nag]   Nagel, C. H.: Untersuchungen über die wichtigsten zum Dreiecke gehöhrigen Kreise. Eine Abhandlung aus dem Gebiete der reinen Geometrie. Leipzig (1836).

[Neu]   Neuberg, J.: Sur une transformation des figures. Nouv. Corr. Math., **4**, 379 (1878).

[OB]   Oakley, C.O., Baker, J.C.: The Morley Trisector Theorem. Amer. Math. Monthly, **85** (9), 737–745 (Nov. 1978).

[Pap]   Pappus: Synagoge (ca. 340).

[Pas]   Pascal, B.: Essai pour les coniques (1640).

[Pon]   Poncelet, J.V.: Traité des propriétés projectives des figures. Mett, Paris (1822).

[Pto]   Ptolemy, C.: Megalé suntaxis tès astronomias (ca. 150), better known as The Almagest.

[RC]   Rouché, E., Comberousse, Ch. de: Traité de Géométrie, I. Gauthier-Villars, Paris (1935).

[Schu]   Schuh, F.: Leerboek der nieuwe vlakke driehoeksmeting. Van Goor, Den Haag (1939).

[Schw]   Schwarz, H.A.: Beweis des Satzes, dass die Kugel kleinere Oberfläche besitz, als jeder andere Körper gleichen Volumens. Gött Nachr., 1–13 (1884).

[Sim]   Simon, K.: Über die Entwicklung der Elementar-Geometrie im XIX Jahrhundert. Teubner, Leipzig (1906).

[Som]   Sommer, J.: Elementare Geometrie vom Standpunkte der neueren Analysis. Encyclopaedie der Mathematischen Wissenschaften, **III/1** (8). Leipzig (1914).

[Ste]   Steiner, J.: Elementare Lösung einer Aufgabe über das ebene und sphärische Dreieck. J. Reine Angew. Math., **28**, 375–379 & Tafel III (1844).

[Ter]   Terquem, O.: Considérations sur le triangle rectiligne. Nouvelles Annales de Mathématiques 1, 196–200 (1842).

[Tho1]   Thomson, W.: Extrait d'un lettre de M. William Thomson à M. Liouville. J. Math. Pures Appl., **10**, 364–367 (1845).

[Tho2]   Thomson, W.: Extrait de deux lettres adressées à M. Liouville. J. Math. Pures Appl., **12**, 256 (1847).

[Too]   Toomer, G.J.: Ptolemy's Almagest. Princeton University Press (1998).

[Vel]   Veldkamp, G. R.: Triangles of reflexion (vertices). Nieuw Tijdschr. Wisk., **73** (4), 143–156 (1986).

[Ver]   Versluys, W.A.: Inleiding tot de nieuwere meetkunde van de driehoek. Amsterdam (1908).

[Vir]   Virgil: The Aeneid (ca. 29–19 BC).

[Viv]   Viviani, V.: De maximis et minimis geometrica divinatio in quintum conicorum Apollonii Pergaei adhuc desideratum. Ioseph Cocchini, Florence (1659).

[Vri1]   Vries, J. de: Over vlakke configuraties waaruit elk punt met twee lijnen incident is. Verslagen en Mededeelingen der Koninklijke Akademie voor Wetenschappen, Afdeeling Natuurkunde, **3** (6) (1889).

[Vri2]   Vries, J. de: Sur les configurations planes dont chaque point supporte deux droites. Rend. Circ. Mat. Palermo, **5**, 221–226 (1891).

[Wal]   Wallace, W.: The Mathematical Repository, 111 (Mar. 1799).

[Wei]   Weitzenböck, R.: Über eine Ungleichung in der Dreiecksgeometrie. Math. Z., **5**, 137–146 (1919).

[Wel]   Wells, D.: The Penguin Dictionary of Curious and Interesting Geometry. Penguin, London, 122–124 (1991).

[Zac1]   Zacharias, M.: Der Caseysche Satz. Jahresber. Deutsch. Math.-Verein., **52** (1942).

[Zac2]   Zacharias, M.: Elementare Geometrie und elementare nicht-Euclidische Geometrie in Synthetischer Behandlung. Encyclopaedie der Mathematischen Wissenschaften, **III/1** (9). Leipzig (1920).

Index